Spring Salmon, Hurry to Me!

Spring Salmon, Hurry to Me!

THE SEASONS OF NATIVE CALIFORNIA

Edited by Margaret Dubin and Kim Hogeland

Santa Clara University, Santa Clara, California
Heyday Books, Berkeley, California

This book was made possible in part by generous grants from the National Endowment for the Arts, the LEF Foundation, and the Book Club of California. Many of the works in this anthology first appeared in *News from Native California* as part of the Four Seasons Project, funded by the National Endowment for the Arts and the LEF Foundation.

© 2008 by Heyday Books. All rights reserved. No portion of this work may be reproduced or transmitted in any form or by any means, electronic or mechanical, including photocopying and recording, or by any information storage or retrieval system, without permission in writing from Heyday Books.

Library of Congress Cataloging-in-Publication Data
Spring salmon, hurry to me! : the seasons of native California / edited by Margaret Dubin and Kim Hogeland.
p. cm. ~ (A California legacy book)
ISBN 978-1-59714-079-9 (pbk. : alk. paper)
1. Indians of North America~California~Social life and customs. 2. Indian philosophy~California. I. Dubin, Margaret Denise. II. Hogeland, Kim.
E78.C15S684 2007
390.089'970794~dc22
2007052030

Original woodcuts by Frank LaPena. Cover art: *Songs of Spring* by Frank LaPena.
Cover and interior design by Rebecca LeGates
Printing and Binding: Thomson-Shore, Dexter, MI

Co-published by Santa Clara University and Heyday Books. Orders, inquiries, and correspondence should be addressed to:
Heyday Books, P. O. Box 9145, Berkeley, CA 94709
(510) 549-3564, Fax (510) 549-1889, www.heydaybooks.com

Heyday Books is committed to preserving ancient forests and natural resources. We elected to print *Spring Salmon, Hurry to Me* on 50% post consumer recycled paper, processed chlorine free. As a result, for this printing, we have saved:

5 Trees (40' tall and 6-8" diameter)
1,976 Gallons of Wastewater
795 Kilowatt Hours of Electricity
218 Pounds of Solid Waste
428 Pounds of Greenhouse Gases

Heyday Books made this paper choice because our printer, Thomson-Shore, Inc., is a member of Green Press Initiative, a nonprofit program dedicated to supporting authors, publishers, and suppliers in their efforts to reduce their use of fiber obtained from endangered forests.

For more information, visit www.greenpressinitiative.org

CONTENTS

DARRYL WILSON Foreword: Seasons, Time, and the "Way" VII

WINTER

FRANK LAPENA *Sleepy Time* 1
KARUK *from "Coyote's Journey"* 3
SHASTA Coyote and the Moons 4
MAIDU The Wolf Makes the Snow Cold 6
YANA *from "Coyote and His Sister"* 7
STEPHEN MEADOWS The Burial 9
JANICE GOULD New Year's Day 10
SYLVIA ROSS Winter Beginning, Winter Ending 13
DEBORAH MIRANDA Petroglyph 19
GEORGIANA VALOYCE-SANCHEZ Winter Comes 21
DARRYL WILSON *Ascui* (When the Ice Cracks) 23
GREG SARRIS Frost 25

SPRING

FRANK LAPENA *Songs of Spring* 29
WINTU *from "Four Dream Cult Songs"* 31
ATSUGEWI The Lazy Man and the Tamciye 32
NOMLAKI Spring Dance 33
DOROTHY RAMON Picking Yucca Flowers 34
SHAUNNA OTEKA MCCOVEY Conception 36
GREG SARRIS Iris 38
STEPHEN MEADOWS In the Mountains 48
DEBORAH MIRANDA Water 49
DARRYL WILSON *Kweme Psukitok*, Spring Maiden;
Amal, Flower Maiden 51

SUMMER

FRANK LAPENA   *The Sound of Insects and Heat*   53
YOKUTS   Rattlesnake Ceremony Song   55
FRANCISCO PATENCIO   *from* "The Quail Legend"   56
DOROTHY RAMON   Why People Did Not Kill Tarantulas   58
DEBORAH MIRANDA   In Praise of August   60
GEORGIANA VALOYCE-SANCHEZ   Summer 1945   61
STEPHEN MEADOWS   In the Water over Stones   64
SHAUNNA OTEKA MCCOVEY   Fireworks   67
ROBERTA CORDERO   Haiku/Senryu   68 ·
GREG SARRIS   Osprey   69

FALL

FRANK LAPENA   *Gathering and Feast Time*   77
ISHI, *translated by* LEANNE HINTON   *from Ishi's Tale of Lizard*   79
LUCY THOMPSON   The Acorn   81
SHASTA   Coyote and Raccoon   83
WINTU   Puimeminmak, the Deer Maker   86
DEBORAH MIRANDA   Faith   89
SYLVIA ROSS   Guns and Roses   91
GEORGIANA VALOYCE-SANCHEZ   The Eye of the Flute   95
JANICE GOULD   U.C. Mascot, 1959   98
SHAUNNA OTEKA MCCOVEY   Embolus   100
DARRYL WILSON   Splashes of Red
Autumn 1867, Tawutlamit Wusci   102
STEPHEN MEADOWS   Alejandro   105

CONTRIBUTOR BIOS   107
PERMISSIONS   111
ABOUT THE EDITORS   118

Foreword BY DARRYL WILSON

Seasons, Time, and the "Way"

The changes in nature commonly called seasons—winter, spring, summer, and fall—were not seasons according to previous generations, they were "times": wintertime, springtime, summertime, and the time to prepare.

Ascui* (when the ice cracks) was a time to live under the snow, dreaming of spring. It was a time when the elders passed on the wisdom they acquired over their lifetime, a time for lessons and legends. It was a time to watch the cold sun move slowly south, a time when spring, fresh life, and fresh air were longed for, a time to grind seeds and nuts, to make flour from dried *apas* (roots), a time to true arrows and to sand the bow, and to plan for the approaching spring. It was the time to soak jerky for the older folks, a time for hunger to be held at bay by chewing dried, salted meat and smoked salmon. The older folks argued about the return of the sun, watching it slowly move south down the frosted horizon, where it lingered before starting its move north, signaling that spring was approaching.

Psukitok (springtime) came with much relief and with many seasons. *Apas* season, flower season, fresh leaf season, blossom season, fish trap building season. There was a season to check the salmon nets for broken strands, a season to run into the landscape searching for food and medicine, a season to study the birds, searching the bush for where their nests might soon be, a season to eat fresh poison oak leaves (medicating the body). Pine season, a time to locate a pine, damaged and repairing itself with pitch (more medication). Spring was a time to sing

---

* Italicized words are in *Aw'te*, the language native to Hat Creek (Atsugewi) and Dixie Valley (Oporegee).

and dance, to gather and eat together, to dream and to dare the dream to come true, a time for older brothers and uncles to teach the youngsters how to survive in the landscape, leading us through the mountains and the valleys.

Apnui (summer) was a relaxing time because it came before the fall, when people worked ceaselessly to prepare to survive the approaching winter. Summer was a time to enjoy life. It was the time for a vacation. Among my people it was a time for ceremony, a time to cleanse the body and the spirit, to think good thoughts, to visit and to entertain visitors and to rejoice.

Nahok, harvest season, came with much work to prepare for survival through the frigid winter. It was a time to get fish, dry and smoke them and layer them in baskets for the time when the sun heads south. It was a serious time regulated by the landscape. Rabbit season, the time when rabbit skin and fur were the thickest and most durable. Hunting season was the time to get deer to salt and dry for the long winter. Salmon *mayukaoho* (last run) was the time to catch salmon, smoke and dry them, also for winter reserves. Dry salmon eggs in the heart of winter made a nice breakfast, and they changed the flavors of many winter foods. Bear season, autumn, was the proper time to get bear because they were fat, preparing for their long hibernation, and the hides could be worked all winter. Fall was a time to plan and lay net traps for the variety of migrating waterfowl. Salted and dried, they were a happy change of the menu while snow smothered the land and ice blocked the rivers and streams. Pine nuts filled the pineapple-appearing cones high in the trees. Climbing to get them was laborious and dangerous, but the flavor-rewards came deep in the winter, too.

In previous generations, life progressed through these times in a regular and balanced way. This way changed one day when strangers entered the homeland with cannons, swords, and angry laws. The teachers were no longer grandmothers and grandfathers, uncles and big brothers, but strangers from strange places. These strangers knew nothing about seasons and cycles and migrations, about hunting, fishing and catching animals and fish for food, and about drying and smoking meat, fish and roots for winter. They knew nothing about gathering seeds and roots or tanning hides. The traditional ways were instantly crushed by strange laws created by strange people. Many of the educated strangers and their descendants are still strangers to the earth and are still here, making laws.

Today the old ways are gone; instead, we have poems and memories about rabbit drives, antelope charming, dancing, singing at dawn and at sunset, mountain climbing, and ceremony. But the poems and memories are not a longing for the past, they are a planning for the future. The "times" and the "ways" have not really changed or disappeared. It is the rest of the world, the people and politics, that has changed, and nature will deal with that. Nature awaits a time when people will come to their senses and live seasonally, in harmony with the earth. These stories and poems are offered in that spirit: to keep alive knowledge and feeling toward the land and the seasonal rhythms by those who have known it the longest and lived with it the most intimately. We must plan for our future, a future in accord with our times and ways, and as a wise man said, we must "conduct ourselves like a great people because we came from great people."

WINTER

KARUK

from "Coyote's Journey"

Young brodiaea plant,
      you must come up quickly,
           hurry to me!
Spring salmon,
      shine upriver quickly,
           hurry to me!
My back has become like a mountain ridge,
     so thin,
         so hungry.

SHASTA

Coyote and the Moons

Long ago, when the first people grew, there were ten Moons. The people gathered together and talked. "Shall we kill the Moons?" said they. "The winters are too long." Coyote was there with them. "Yes!" said he. "I am the one who can kill them. I will do it." The Moons lived far to the eastwards. A great bird called Toruk lived there too. The Moons had taken out his leg-bones, so he could not go away. Every day they went to gather roots and left Toruk in the house to guard it. He cried all day. When he was hungry, one of the Moons went and fed him. Every night they brought back roots. One came bringing big snowflakes with him as he came; one came with a shower of rain; one brought great hail; one brought strong winds, so that great trees were blown over…The other five were not as strong.

    The people said to Coyote, "Well, you go." So he went. "I will fool them well," said Coyote. The people told him what to do. He went to where the Moons were. He went to kill them. When he got close, he found they were gone gathering roots. Toruk was there alone. He was frightened. He almost called out in warning. "Be still, Uncle! It is a friend," said Coyote. "Here is food for you. Eat it. I will fix your legs for you." Toruk had no legs, for the Moons had taken out his leg-bones. Coyote fixed Toruk's legs. He cut up some young black-oak, and made legs out of that.

    "What do they do for you?" said Coyote. "When I am hungry, I cry, and one of them brings me food. That is what I do," said Toruk. "Good!" said Coyote. "Do you cry out now, and a Moon will come." So he cried out, "Tō-ō-ō!" Then the Moons said far away, "Ha! He is hungry. Do you go and take him some food."—"Very well," said one, and he went. "He is coming!" Toruk said. Then the storm came, it poured down. Coyote slipped behind the door, and watched for Moon when he should come in. Soon Moon came; and when he put his head in the door, Coyote cut it off. He seized him by the hair, and cut off his head. Then he threw the head behind the door, and the body to the other side of the house. Then he warmed his hands by the fire, and got warm again. "Now cry again!" he said to Toruk. "All right!" said he, and cried, "Tō-ō-ō!"—"Oh! the slave is not satisfied," said the Moons; "I guess you had better go."—"All right!" said one of them. "He is coming!" said Toruk to Coyote. So the second Moon came to the house; and as he came in, Coyote seized him by the hair as he stooped, and cut off his head. He did then as before, threw the head back of the fireplace and tossed the body to one side.

He was nearly frozen, he warmed his hands. When he was again warm, he said, "Cry again!" The Toruk called, "Tō-ō-ō!"—"Ah! what is the matter with that slave?" said the Moons. "He is calling again. You had better go." So they said to the biggest Moon. "All right!" said he, "I don't know what is the matter with him," and he went. Then Toruk said, "Here comes the biggest Moon!" Coyote was nearly frozen stiff, it was so cold; everything froze, everything cracked. When the Moon put his head in the door, however, Coyote did the same as before, seizing him by the hair, and cutting off his head. Coyote was almost frozen to death, he was numb…

Now he had killed five Moons. Then they found out what was the trouble. Now, Toruk said, "They have found out what has happened. The last one that was killed got his hair in the edge of the fire. They have smelled the hair burning, out there where they are picking. Let us run away!" So Coyote and Toruk ran, and got away. If Coyote had not done this, there would have been ten Moons. Coyote killed five of them.

MAIDU

The Wolf Makes the Snow Cold

Wolf and his wife lived toward the southwest. They had a daughter, who was married and had many children. The children were out playing, when it began to snow. It kept snowing till the snow was up to people's knees. Then it cleared off. Next morning the children went out and began to play. They made a great deal of noise, shouting and calling to each other, as they played in front of their grandfather's house. The children played all day, and next morning they began again. Toward night the old Wolf grew angry. He wanted to sleep, but the children kept him awake. It was the first time the children had ever seen snow, that was why they made so much noise. Wolf said to his wife, "I will teach those children something." Then he went outside the house, and urinated in the snow, all about the camp. That made the snow cold: before, it had been warm. The children played about awhile; but their fingers and toes soon got cold, and they went into their mother's house to warm themselves, and cried. Then Wolf went back into the house, and went to sleep. That is the way he spoiled the snow.

YANA

from "Coyote and His Sister"

Coyote went north and turned east, leaving Clover creek to the north. He went east to Bagatᵋdidja'myak!aina, that far he went. Coming up from the west, Coyote had an otter-skin quiver, and very good was the flint in his quiver. He had white feathers and put them into a net-cap, an eagle's white breast and leg feathers he put into the net-cap. Coyote did not have merely arrow shafts put under his arm, these were all provided with flint arrowheads. Frost came from the east. Frost also had a net-cap filled with white feathers, he had his feathers made of snow. Very pretty were Frost's white feathers. Frost was going west, Coyote was going east; they met each other at Ganu'myā. "Hᵘ!" panted Coyote. Coyote sat down, Frost sat down. "Whither are you going?" asked Coyote. "I am going west," said Frost. "Indeed! I am going east," said Coyote. "Indeed!" said Frost. "Tell me," said Coyote, "how are the east people getting along?" "There are no people. I did not see any," said Frost. "Hê! Very beautiful are your bow and your arrows. Hehé!" Coyote said, "I should like to have your white feathers," but Frost said nothing. "Let us change about," (said Coyote). "This bow of mine is bad, these arrows of mine and my white feathers are bad." "Oh, well! Let us change about." "Yes," said Frost, and he gave him his arrows, his net-cap filled with white feathers and his bow. "Let us trade good things with each other." Frost handed his net-cap filled with white feathers to Coyote. Now Coyote put white feathers made of snow on his head; just so Frost put Coyote's white feathers on his head. "Well!" said Coyote, "I am going east. Do you for your part go west."

Now he went east, while Frost on his part went west; now they departed from each other. Frost laughed. Coyote went east, and (soon) said to himself, "I am sweating." Really it was snow that was melting, the water came dripping down on Coyote's face. He looked back at his bow, he looked back at his flints and arrows. No arrows were to be seen, no bow was to be seen, they had all melted away. Coyote stood there and looked all around; Frost had gone far off to the west and was no more to be seen. Coyote put his hand on his head, felt around on his head for his white feathers, but the white feathers were no more. Coyote stood still, pondering. "Dam ͝ nimā'na!" said Coyote, "you had good sense, young Frost! I thought indeed they were real white feathers," said Coyote. "That is why I changed about with you. You had good sense." He went on east with nothing now, without bow and without white feathers. Frost's white feathers did not melt, nor his bow and arrows. Coyote now went off home, until he arrived at Ha'udulilmauna.

STEPHEN MEADOWS

The Burial

A scene one remembers
from a night full of dreams
The wet pit disguised
by a square of green cloth
Here and there small birds
forage in the grass
pecking to and fro
for each sad urgent seed
Beyond the white sand hills
comes the hard boom
of the rip surf
clubbing winter beach
Small fists of relatives
chat soft among the markers
relieved it is over
they are patterned like beads
about the nondescript stones
and the casket
incidental in the cold

JANICE GOULD

New Year's Day

1.
An edge of despair shimmered
like light on the ocean,
just before the darkest storm hit
land, just before the pelting
rain swept the far Aleutians,
the cold Pacific. When
the black cloud raced in,
whistling, shrieking, we were
giddy, excited. Surf glimmered;
the squall gathered strength,
pressed the coast, pushed
at the shore. We struggled
with the wind, leaned
against it, shoved
our shoulders into it,
defied the way it cut
our faces, brought tears
to our eyes. We laughed
anyway at the roaring,
the incessant tempest
that rushed over chaparral,
twisted the poison oak, the sticky
Monkey Flower. What bird
could live through this without
stumbling, without being torn,
bone to feather, severed,

New Year's Day

scattered? The light was
less friendly than the cloud,
cloud less hostile than
the wind; wind battered

the stubborn earth that clasped us,
pulled us down, pried
open our beaks
against the dirt, the mud.

2.
Piled in the car, my family
was ruddy with cold,
noses dripping, eyes tearing
from the wind. The rain came
pounding out of the charred
clouds, slapped the roof of the car.
We passed through Pt. Reyes
farmland, the dairy operations.
Animals with muddy legs,
filthy bellies and udders,
stood in sodden pastures
that would not dry till August.
Meanwhile rain ran in rivulets
down dark cow paths, soaked
into the roots of beach grass, lupine,
beneath the stalks of dead thistle.
Further inland, creeks
the color of powdery chocolate
churned between banks
that shone with red clay.
Did we sing? Were we
silent? We took roads
where mists tangled
in boughs of apple trees
planted years ago by Italian farmers.
My brother-in-law brought out

a box of crackers. We opened
a thermos of tea, and the three
in the backseat steamed the windows
with their chatter.
The smell of damp wool
rose from caps and jackets.
Wedged in the front seat
between Dad and me, Mama
took my hand, tapped it
against her knee. I marveled
at how small she'd become,
her warm body, her laughter.
In February, she would discover
the cancer, a protrusion
beneath the skin. For once
she was at peace
with her daughters.

SYLVIA ROSS

Winter Beginning, Winter Ending

At the beginning of winter, 1919, when it was finally cold enough in the valley to keep meat hanging in the shed from spoiling too soon, a man named Ernie took it into his mind to go up into the hill country of Madera County to his wife's people. His plan was to get venison, even perhaps bear meat to season and smoke for winter. He favored elk, but the elk herds were already disappearing. He hadn't seen an elk in three years of trekking the Sierra with his wife's people. Ernie liked his wife's relatives. Indian people didn't seem to mind the scarred face or blind white eye that he presented. They took him for what he did, not for how he looked. They were the truest people he knew. He packed bags of raisins the Italians down the road had given him to take up to Mrs. Billy. He knew she'd like that, and not mind Billy going off with him to hunt for a few days or a week. This year Sarah Ellen might get her first deer. That'd be better than elk or bear.

Ernie was a white man and his wife, Reenie, had been dead for nearly seven years. He'd promised her as she was dying that their daughter would know who she was. He kept his promise and his daughter Sarah Ellen knew who she was. She knew her mother had been Chukchansi and therefore she was too. She knew that her mother had been born up in a place called Fresno Flats. Ernie told her that her mother had signed on as a harvest crew cook in 1909, 1910 and 1911 to save the money to buy the forty acres where they lived and he had taken care of her when her mother was away.

The little girl knew that if her mother hadn't died, she'd have a little brother six or seven years old and a mother like everyone else. But since she didn't remember them, she didn't miss them.

The best part of her life was going up into the hill country. Ever since the little girl could remember, she had traveled in the buckboard with her dad up into the mountains six or eight times a year. She liked the long ride across the plains, and sleeping out under the night sky. Ernie would tell her of her mother and her grandmother. He told her most of the Indian people in the hills were her mother's relatives and therefore they were her relatives. She thought often of her grandmother. Her grandmother had been dead long before she was born. But, she knew that her grandmother had been taken away from her tribe to be a worker on a ranch in Crane Valley when she was a tiny girl, and then when she was almost grown had run away from that mean ranch to the stage depot at Fort Miller, where she got paid real wages for kitchen work. Sarah Ellen thought that was a very brave thing to do. She couldn't imagine running away from what she knew to what she didn't know.

She liked it when they went up by the reservoir that covered Crane Valley with water. Ernie told her the water had filled it up in 1912, that both Uncle Billy and Tom worked on the dam. She thought about what had happened to the ranches that used to be there. Were there still houses and barns and fences under the water? She wondered and wondered. The long wedge of blue water behind the dam taught her that everything changes and nothing ever stays the same. Once she had a mother and now she didn't. Once there was a ranch where a stolen Indian child worked and now there was a shimmery lake. Once the digger Indians were naked savages. It said so in her schoolbooks. But now Uncle Billy wore plaid shirts and overalls just like her dad. She wondered if she were a digger Indian. She thought maybe she was.

She was never frightened of traveling up into the mountains. Her dad was always with her and everyone she met was good to her. She liked the old men who'd fish and hunt with them, her Indian family. There was only one woman she knew. Mrs. Billy. She never left her cabin and didn't speak good English, but she was always happy to see Sarah Ellen. Mrs. Billy called her "Little Reenie." But this trip was impulsive. She hadn't known it was coming. She'd felt headachy walking home from school that Wednesday in early December and thought she would lie down on her cot until time for supper. But when she came to their lane Ernie was busy loading up the wagon with supplies. A barrel of stock grain and two cans of kerosene were in the bed of the wagon, and her dad was lifting the heavy canvas tent over its side. She didn't feel her usual excitement. She didn't feel really good. Besides, she was in fifth grade now and didn't want to miss school. Teacher didn't like it when she was absent.

Her dad had given her a big smile when he saw her come up the lane through the fog. He told her to get her bedroll ready, oil her snake boots, and get her warm mountain gear. He reminded her that there might be snow if they'd need to climb way high to get to a good spot to find deer or bear.

She told her dad that teacher wanted her in school and he made a face at her. "I don't want to miss school," she told her dad. "I can stay with the Dellachios and go to school every day."

"You have lots of time for school, Cheeri Biri Bin," Ernie answered. "We are going to climb up and see the sun! You spend any more time with those Italians, you'll forget who you are."

She told her father she didn't feel well. But Ernie just put his hand on her forehead to feel for fever. He shook his head and told her that she "…just had the pip."

"Child o' Mine," he said, "If you are too sick to hunt with me and Uncle Billy and Tom, you are too sick to go to school." She saw his logic. Yes, school would keep.

The next morning, they were up before dawn. The horses hitched up easy and two of them drove east through the cold misty fog toward the mountains. She felt better. It was cold but she had a blanket wrapped around her and wool socks on her feet. Ernie was singing softly as Sarah Ellen watched the light come to the ranches and farms that they passed. She watched for trucks and automobiles. She wished her dad would get an auto and drive her to school in fine style. Four of the children in her class had either an auto or a truck from Mr. Ford. She knew Mr. Packard made autos too. They didn't leave poop everywhere and she knew that had to be a fine thing for the world. The cool winter daylight made the road ahead of them grow brighter and as they traveled east the fog began to fade. Her father had a pretty little rifle he'd bought from a man down at La Vina store. She wondered if he'd let her shoot it on this trip. He wouldn't let her shoot his shotgun and his big rifle was too heavy. She never could hold it steady. Uncle Billy had a lighter rifle and with it she was a good shot. Better than her dad. But then, she could see better because both her eyes worked.

It was dark by the time they got to Uncle Billy's cabin. Mrs. Billy came out to the porch and took Sarah Ellen inside. She gave her a supper of sour berry mush while the men outside roasted rabbit. The dried berries made the mush tangy and good and the little girl usually liked them, but this night the berries made her mouth hurt. She didn't eat much, but rolled out her bedding where she usually slept over by the woodstove and went to sleep.

In the morning Uncle Billy, Tom, and another man Sarah had never met packed the horses and began the long trek to the deer grounds. The other man, Solo, used a different kind of knot when he cinched up the pack animals. Tom told her Solo was a Mono, not a Chukchansi, and they did things different. Mrs. Billy had given her a little bag of jerky and she had her canteen full of mountain water. Walking beside the horses, she felt better. When the sun was high they stopped and she could look down and see the lake on top of Crane Valley. They ate lunch, made coffee, and rested awhile. Sarah Ellen wanted to go to sleep, but it was time to go on. The trail got steeper and now and then one of the horses would misstep and make a clunky noise as its hoof would slip on a rock or some gravel. It was warmer, and she took off her jacket. Late in the afternoon her headache came back. But it was just the pip, so she didn't say anything.

They didn't see one deer all day. No scat either. In the evening the men took turns singing. Tom and Solo tried to outdo each other with their Indian songs and Solo made a soft pounding noise on the back of a frying pan with a chamois over his hand. His drumming made Sarah Ellen want to step dance around the campfire as she had on other hunting trips, but she was too tired. She cuddled up close to her dad, and he took his turn at singing army songs. The men got quiet. Ernie sang "Apple Blossom Time." Then he put his arm around her and hugged her. He sang her favorite song, "Two Little Girls in Blue," and Sarah Ellen was happy there in her mother's own mountains listening to her father's song. She climbed into her bedroll. She could hear the men talk. Around the fire in the winter night the men talked about setting up a business to pack easterners into the mountains now that the war was over. No place would give better deer or trout than here. "Better than working lumber," Tom said. "Better than working the ranches," Solo countered.

The next day Sarah Ellen woke up feeling sick and her headache was back. But she said nothing and that day they climbed higher. She walked slower with her dad while Uncle Billy, Tom, and Solo led the horses up the trail ahead of them. Ernie told her how the Chukchansi and the Mono Indians used to be enemies. "See how Tom and Solo are friends? One day the Hun and the Americans will be friends again."

She knew he was talking about the war that was just over. When they were in a stand of white pines they found deer trace. Solo pointed off to the left and Uncle Billy nodded.

They turned a mile in that direction and then tied up the stock. Tom and Solo went off with their guns but came back with no kill but a couple of birds. By that time Uncle Billy and Ernie had set up the tent and made a fire. Supper was cooking.

That evening Sarah Ellen wasn't hungry and she felt hot. She went to bed early and didn't even want to listen to the men talk. Her face hurt and her stomach hurt.

She could smell her dad's pipe smoke blending its smell with the campfire. She thought she could smell sweet grass burning and hear Uncle Billy praying for a good hunt.

She was finally lulled into a restless sleep. She woke up in the middle of the night and knew she was sick. Then she didn't know anything more until a doctor at the Burnett Sanitorium in Fresno sat down in a chair by her white bed in a white room and told her she was lucky to be alive.

"Those Indian men saved you, little girl. They forced some kind of Indian tea down your throat and used snow packed in blankets to pull your fever down. Why, your father told us that two of those men ran for miles to a mountain to bring snow down to cool you. You'd had fever fits in the night. Once, they thought you were dead. The men built a litter and toted you down out of those mountains without stopping to eat or rest."

She found she'd been in the hospital for two weeks. She asked the nurses what was wrong with her and was told that she'd had the worst case of mumps any of them had ever seen. "Only diphtheria and

Spanish flu brings children as near to death as you have been, Dearie," one of the white-capped nurses told her.

Her dad came in and sat by her bed. He looked sadder than she had ever seen him. "Didn't we get a deer?" she asked him. He just put his head down on the bed and she could see he was crying. Everything in her life changed that winter. The public health service reported that a little girl, eleven years old, had nearly died while camped out in the wild with four men. Affidavits were signed. There was public outrage and Sarah Ellen was taken from her father's custody.

Sarah Ellen went to live in a succession of foster homes. Eventually she graduated from Madera High School while living with a Seventh-Day Adventist family. She never went hunting again. Nor did she ever eat sour berries or look down from a mountain to see the lake that covered Crane Valley. But Ernie had kept his promise to his wife. Their daughter never forgot who she was.

I am her daughter. The first thing she taught me was that I was a Chukchansi child. The second thing she taught me was that everything changes.

DEBORAH MIRANDA

Petroglyph

Snow falls that night,
spreads heavy and smooth
like stone, like white granite.
It takes the sharp cut of deer tracks.

In nightgown and bare feet,
she follows a string
of cloven hearts wandering from the woods,
past the barn with its scents of straw,

cats, cobwebs; lapping the length
of the skinny tin trailer
where the girl had lain curled
in dreams of slow words; past

her father's red truck
asleep in the driveway, dents filling with snow,
tools covered in the bed made
fresh and clean, no trace

of labor, his sweat, jumbled scraps of lumber; down
the long driveway, to enter mute pines
and bare maples at the mouth of the road
that leads away.

WINTER

She stands breathing in silvered swirls, heart
thumping; *this is as far as I go*. Snow
takes her print, curved half-moons
cut by the heat of childhood in skin.

GEORGIANA VALOYCE-SANCHEZ

Winter Comes

Winter comes

The sun is low across
the sky   gray days
daily struggles
and discord   weigh
heavy
like fallen snow
on tree limbs
bent near to breaking

   but we do not break

Winter comes

Old Winter Songs call
carried on the chill
night wind
love songs to the deer
Prayers
for the good hunt
for baskets of pinon
and acorns   wild rice
and corn   overflowing
Prayers
for the children
children

who cannot know
of the harsh winters
to come   children
tumbling in play
like bear cubs
in springtime
and oh the stories
we tell
of who we were and
who we are
the Sacred Stories
that give us
life

Winter comes

Time for all clans
to work as one
clan   one people
one   offering
to the Creator
to halt the sun's descent
once more
we stand   together
like worn prayer sticks
eagle feathers fluttering
in the wind
at early dawn
remembering
once more
that winter always comes
with Winter Solstice
and rebirth
of the sun's ascent
across the endless sky

DARRYL WILSON

Ascui (When the Ice Cracks)

It was a few days after Christmas, 1957. I made presents for no one, no one made presents for me. My military training complete, I was preparing my hesitating body and troubled spirit to "move out" to my next assignment, Okinawa.

The hesitation and troubling were caused by the "new laws for a new time" being promoted by the Bureau of Indian Affairs that arrived in the homeland in the form of the 1934 Indian Reorganization Act, authored by Secretary of the Interior John Collier. The icy ramifications of that Act were discussed by the council a week earlier. The council saw it as a collection of word trickery designed like an avalanche to smother any remaining native cultural rights and rules, from sea to shining sea.

Lasiki (daybreak) my restless spirit made me crunch through the crusted, frozen snow, making munching sounds as my boots mashed the sugary substance below. We struggled under a bright half-moon that seemed suspended in the eternal, crystal landscape. Breathing little puffs, we looked east. Frosted mountain ripples whispered of a rugged horizon. To the west, moonlight splashed ghost-silver across the homeland, casting black, brittle shadows. *Ume Juma* (River of Stars) moving like an ice clogged canal, oozed slowly and laboriously into the frozen distance. Early sun lanced the eastern horizon, crowning the higher peaks in shivering sterling.

In the hush across the arctic landscape, a lonesome coyote called with a trembling voice. Shivering moon slid iceberg-slow across the sparkling, frosted sky. *Sukahow* feathered the early darkness on silent wings, a fleeting shadow upon the brilliant snow, hunting the frigid ridges.

My worries about the native future hung like icicles from my mind and shivered while responding. Like a hibernating winter rattler, my chilled thoughts moved very slowly, very slowly.

"There must be a future for our children or like frost-bitten leaves on a winter oak, the indigenous spirit will atrophy, fall to earth, and be carried away with a blast of frigid wind. But what can it be?"

My "bothers" hung like icicles from my mind, freezing sharp and to a point before they fully formulated. My spirit took me by the hand and we crunched back to the cold pickup, breathing frost-puffs in the moonlit, harsh daybreak.

GREG SARRIS

Frost

Winter.
Not ancient stories about the time, before this one, when the animals were still people, before Coyote messed things up with his hapless machinations. Nor the dark room, warm but still black as the cold mid-winter night outside, with nothing but the floating voice of the storyteller impersonating the people in the stories: crafty Coyote's devious whispering; Blue Jay's shrill admonishments; Frog's old-man rasp; Quail, the most beautiful of all the people, her gentle-as-brook-water songs. None of those things. But cows—feeding the cows, their cloven hooves planted in the frost-covered earth, nostrils blowing steam above an unfastened bale of alfalfa.

I would stand, warming my hands in my coat pockets, for hours watching the cows. I had a good eye then, not just for an old cow's swollen knee, or maybe a rheumy eye, but even the faintest rise in her hide, indicating the presence of a grub. And the alfalfa, too: it was best if you could see dried purple flowers, sign of an early June harvest, after just enough warm weather. I was seven. I had no cows of my own, but followed the local dairyman. I wanted a cow.

No Indians at home either. I was adopted. At the time, I knew nothing of my birth father, Emilio Hilario, or my Coast Miwok heritage. That would come twenty years down the road. And I would hear about the things the old-timers did. Winter activities; storytelling, for instance. Renowned Pomo basketweaver and doctor Mabel McKay, who I was fortunate to have known, explained the rules about the ancient-time stories: "Only tell them in winter, after the first frost and before the last frost. Think about them then, their meanings. Not in summer, when there's snakes and things in the grass and you need to pay attention to where you are going." But that has nothing to do with memory, what surfaces from experience, as I recall winter now.

There was a man named Tommy Baca. He had only one arm, and he was a house painter. My mother said a thousand times no one could mix color like Tommy Baca. He was a stocky man of medium height, with a broad handsome face. He had thick, wavy black hair. He smiled a lot. I marveled at how he kept papers and such tucked against his side, just below his armpit, with the stub of his missing arm, and the way the stub would move, seemingly of its own volition, when he was excited, though I was careful not to let him find me looking. "Don't stare at people," my mother snapped. He was Indian, Coast Miwok; if I knew as much then I don't remember, certainly not the Coast Miwok part. What interested me was that he had cows.

And because my parents were friends with him, I had access to the cows. They were steers, actually; mixed-blood dairy calves, Guernsey and Hereford, which in those days you could get for a drop in the bucket, as the dairy farmers kept only their purebred heifers. Out at Tommy's place, west of town, I could spend hours with the critters. Once his son, Mark, about seven like me, asked, "Why do you always look at Dad's cows?" and I felt as if it was a bad thing, like when other kids called me "adopted."

I'm not clear about this next part. What happened exactly. Maybe because for a month or more I was on cloud nine. Did I hear my parents and Tommy Baca talking in the kitchen after work one night, did I hear about it that way, before the afternoon out at Tommy's when he said, "Pick out the one you want. Take your pick"?

A Guernsey and Hereford cross steer. I kept him in the field adjacent my parents' house. I named him Harry. Why I picked him from a lot of half a dozen others exactly like him, I don't know, or remember. He proved intelligent. He knew the time of day, precisely, when I arrived home from school, and never failed to stand just inside the aluminum gate—always the same spot—waiting for me. He learned to genuflect, lowering himself on one knee, so that you could scratch the crown of his head, and later, when he grew, to make it easier to climb aboard his back. Yes, we rode him—every kid from the neighborhood in those days has a picture of himself or herself astride Harry. His horns grew, but he was gentle, careful even, when he tossed his head to shoo flies or strew a flake of hay over the ground. Except for the time, head lowered, he charged Alan Chaney, chasing him out of the field. But no one cared since everyone knew Alan was "a bully" and had no doubt provoked Harry.

    That was so long ago. A million stories ago. Of course, I found out who my father was—I can look back and understand things I hadn't the faintest idea of before. The connections between me and Tommy Baca, for instance. His grandmother, Maria Copa, sang for my great-great grandfather, Tom Smith, as he doctored the sick. I can imagine them along the coast and through inland valleys, traveling in a wagon and later in a Model T Ford. That box with the stuff of history spills—the ancient villages fall out: Olumpali and Alaguali for Maria Copa, and Petaluma and Olemitcha for Tom Smith; then the Spanish galleons and English steamers, oceans crossed, wars, marriages, priests and soldiers, adobe and brick, overalls and Panama hats, and, still, clamshell disc beads and flicker feathers—and I see in that chance meeting of an adopted boy and a one-armed house painter the miraculous web that is all of time, nothing more, nothing less, all-inclusive. But it's memory that prevails still. Memory trounces this miraculous web, that is, if memory is not the vantage point from which I gaze upon it. No, not even Grandpa Tom's songs left on wax cylinders, or his ancient-time stories left in a graduate student's dissertation; not beads and feathers. Winter. It's feeding the cows—feeding Harry. And frost. The earth is blanketed with frost. A quarter-inch thick at least, on the bare tree limbs, on rocks. Harry too is covered, topped as if with a layer of frosting. Harry steps into the sun to munch the alfalfa I just tossed over the fence, and I see the frost, so cold, so powerful, begin to fall from his back, barely perceptible, trifling dust. And, again, I'm not sure what happened exactly—whether I had overheard "fattened" and "spring grass" in a conversation between my parents and Tommy Baca the night before, or months before and didn't understand or chose not to—but at that moment I got it, understood the whole story, what the words would mean after the last frost, come summer. I might protest but it would do no good. Never mind the purple blooms in the alfalfa. What a complicated and frightening world replaces winter.

SPRING

WINTU

from "Four Dream Cult Songs"

Above where the minnow maiden sleeps at rest
    The flowers droop,
      The flowers rise again.
Above where the minnow maiden sleeps at rest
    The flowers droop,
      The flowers rise again.
Above where the minnow maiden sleeps at rest
    The flowers droop,
      The flowers rise again.

ATSUGEWI

The Lazy Man and the Tamciye

One spring a lazy man went to the bench near Lost Creek to get pine nuts. Here he met two tamciye women who asked him what he was doing.* They said, "We don't eat that kind of food. You better go back with us." They took him to their house on the west side of Bald Mountain, where there was an earth lodge. He saw all the tamciye people. The men were out hunting deer or fishing. In the evening they returned. The lazy man stayed with the tamciye for a while, and they treated him with much hospitality. When he wanted to return home, the two tamciye women made ready to go with him as his wives. He was given a buckskin shirt, pants, a bow and arrow, and other things. Before this he had been naked. The two women loaded themselves with dry meat and other things, and the three started to his home. But before they reached his home he made an excuse to go into the bushes, saying that he wanted to urinate, and as soon as he was out of sight he started running. He wanted to leave the women. They saw him running and were angry. They took back all the clothes and beads they had given him, so that he had nothing on when he arrived home. Then the two women returned to their own home. The man told his family what had happened. They were very angry with him for running away from the two women. He was a fool. If he had brought them back he would have made a good living.

---

* Tamciye are usually friendly little people living much as Indians themselves; they were thought of as guardian spirits who brought luck and strength. Though they seldom interacted with the Indians, favored individuals were sometimes able to see and talk to them.

NOMLAKI

Spring Dance

Our people dance every spring and have all kinds of dances. When everything is all very green, when winter is over and everything is warm and the sun is coming north, then the birds holler witwitwit, and the people begin to ask, "Why can't we play a little?" Then they send news to their close neighbors that they are going to have a dance. The people ask the chief before they can give the dance, and if he agrees he will get out early in the morning and tell all the people to go hunting, fishing, and so on. He will name off the people for the various jobs early in the morning while they are outside listening.

The meat is brought to the chief, who gives each person some to be brought back prepared in the evening. Everybody gets some. At about five in the evening they start to dance, perhaps practicing first—some old fellows with the youngsters. This is a spring dance, a play dance, a home dance. They don't gamble. It's held inside the sweat house. There aren't an awful lot of dances among our people [i.e., there is not a great variety of dance forms]. One is a fast dance called *to-totcono* that takes place in the sweat house.

The dance group included two drummers, two singers, one person to call the dances, from two to six male dancers, and as many female dancers as were available. They dressed behind the drum; the singers stood in front of it, and the timekeeper in front of them with a split-stick rattle in his hands. There is a leader who "dances the girls out" of the dressing place one at a time toward the end of the dance [apparently dancing back and forth from the curtain to his position]. This is said to be a strenuous part. The women sing a sort of chorus in accompaniment to the two singers and then dance back with a heavy beat [stamp of the foot?] behind the drum, where the curtain hides them.

DOROTHY RAMON

Picking Yucca Flowers

Pata' 'umuch 'ashee' 'shee'.   Yaameva' shee'.
The yucca flowers would bloom. They would bloom in the spring.

'Ami' Taaqtam 'amaym chaway mutu' peshkow.   'Amaym qwa'i'.
And the Indians would gather them while they were still
 budding. They would eat them.

'Amaym 'uviht peenyu' qwahow tum hiit pana' nyaawnk chaway.
When they (the yucca blossoms) were ripe they would gather
 them.

Mutu' rrow'nka'tim qwa'i'.   Mutu' 'apanim qwa'i'.
They would eat them while they were still green. They would eat
 it while it's still fresh.

Puchuk 'aqwahi'tti' qaym qwai'i.   'Apenki'tti' qaym qwa'i'.
They don't eat them when they're ripe. They don't eat them
 once they've bloomed.

Qaym 'aqwahi' hiiti' qwa'i'.
They would not eat them when they were overripe.

Mutu' rrow'nka'tim qwa'i'.
They would eat them while they were still green.

Picking Yucca Flowers

Pana'm puuyu' nyihay Taaqtam 'ip terav 'amay. Peepavuha' wuuwert wen.
That's what the Indians used to do on this land. They had a lot of plants.

Puuyn'ayeewpat paavuhac raaqwc tum hiit weney.
There were different kinds of edible plants.

Tum haym mermerher', qay' taaqtam, warrêngk huwam xhinyim 'amay' 'ingkwa' pichim, 'ipim qac terav 'amay'.
Nowadays there are all kinds of people on this earth (non-Indians), who now live here on this continent.

'Aam 'aame'e': peepavuha' wen 'ama' puuyu'ayeewpat waha'. 'Ama' 'ayee'.
And they also have their own kinds of plants. That's all.

SHAUNNA OTEKA MCCOVEY

Conception

It isn't hard to imagine
you were conceived here
in this beautiful place of

canyon walls covered thick
with maidenhair fern,
a stream carving its way

to the enormity of the Pacific.
What a story to tell, that
spring when you became

more than a wish, more
than a dream that
lovers dream.

I wonder where it was
they stole away, the
secluded spot on a soft

needle bed of spruce and redwood,
where the world opened its arms
to welcome you, and

the ocean waves
met the shore with
crashing applause.

Conception

It was a privilege
to take you back
to the place it all began,

to stand with you
in hope and anticipation of
beginnings yet to come.

GREG SARRIS

Iris

Spring.
Mabel McKay, renowned Pomo basketweaver and doctor, she told me about this, too. Spring. "Coming out time," she said. Which was how the season was described, quite literally, by many Native California cultures. New growth, blossoms, sedge sprouting on creek banks—when, after winter, it is no longer safe to tell stories, not only because you must pay attention to where you are going, watchful for snakes and such, but because you too are coming out, becoming story. Living again; living new. Tribes had ceremonies to mark the season. Often every plant and tree was named, every creature even, lest the people forget it, and it, in turn, forget the people. Mabel recalled the *sectu,* or ceremonial leader, in Colusa standing atop the roundhouse entrance at dawn one spring morning, facing east, announcing each part of Creation, as if in that faint light the world itself was emerging for the first time.

   I remember Linda (not her real name). Spring, the miracle of continuation; yes, it's Linda I see. She had relatives on the Kashaya Pomo Reservation, and she took me there once to the Strawberry Festival, a ceremony to dedicate the new fruits. Which was where I first saw Essie Parrish, the great Kashaya Pomo prophet, a big woman in ceremonial dress, traditional long skirt and clamshell disc bead necklace—praying in her language before a table crowded with food, least of which the tin pots and Indian baskets heaped full of bright red strawberries. "Something about spring, Indian things," Linda attempted to translate, though she didn't know the language either. For the Kashaya Pomo, the wild strawberry was the first fruit of the New Year, and therefore became symbolic of spring. The ceremony—the costumes, songs, four nights of dancing that preceded the feast—came from Essie Parrish's Dream. Did Linda know that? I certainly didn't.

That was forty years ago, at least. I didn't know anything then. That I was Indian. That Linda was my cousin. I know about my Coast Miwok heritage now. It's been a long journey—I'm chairman of my tribe, the Federated Indians of Graton Rancheria. We are Coast Miwok and Southern Pomo, descendants of a handful of survivors from Marin and southern Sonoma Counties. We struggle to start anew a spring tradition. We offer prayers in languages we are re-learning, listen to songs found on wax cylinders in museums and university libraries. I've learned some history. And always there's Mabel, deceased, whose advice I nonetheless never forget. Yet it's Linda I recall—and not because she took me to the Strawberry Festival, my first so-called Indian ceremony. Frankly, I didn't think much about it. No, what comes out new, magic to the eye, is a story and it starts with Linda.

Simple enough—I had a crush on her. I'd see her in first hall, not far from my locker, talking with her sister, friends. Heaps of black hair. Eyeliner and lipstick, a mole planted with a brow pencil above the corner of her mouth. She was a woman. She wore tight black skirts—girls had to wear dresses to school in those days—and colorful silky blouses. She offered her friends only a meager nod, a half-tilt of the chin, the tough-girl greeting—no phony smiles, no popular-girl routine. I didn't have a chance. I wasn't a tough guy. Anyone remember them—the guys in skin-tight 401s and white T-shirts, one sleeve rolled up holding a soft pack of Camels above a bulging and, if *really* a tough guy, tattooed bicep? I was a late bloomer: flat-limbed and soft-faced; fourteen and I looked like a twelve-year-old, or worse, a girl, sporting an oily pompadour. I lived in a middle-class neighborhood. I was white—everyone knew the toughest guys were Indian or Mexican. I got the nod one day, but what to say? Ask if she wanted a cigarette? Did I have any cigarettes?

Then some luck. Ritchie, the guy who sat in front of me in math, was Linda's cousin. But it wasn't Ritchie who afforded me an introduction. It was Tommy Baca, the one-armed house painter who had given me, years before, the crossbreed steer I named Harry—Tommy Baca was Ritchie's uncle, and Linda's, which I learned from Ritchie in an unrelated conversation.

I told her I knew her cousin. She narrowed her eyes, sizing me up I figured. Had she really given me a nod, or had I imagined as much? My stomach was in knots. What if one of her friends came along and caught her talking to this gangly kid? Would she say something to humiliate me?

"Ritchie?" she finally said. Her mouth barely moved, as if even the name of her cousin she intended to keep to herself.

I told her no and mentioned Tommy's son, which caught her off guard, then launched into a pathetic story about Harry, who ended up on dinner plates in the "Saddle and Sirloin."

She offered the faintest smile. At lunch that day she handed me a note folded eight times into a hard square: "You are my friend. L."

We walked the halls after that, sat together at lunchtime. I learned who she liked and who she didn't. Which girls she wanted to "kick their asses." I met her sister, her friends. I met tough people, the *really* tough people. Everything changed. People looked at me differently. The hoods noticed me. Teachers glared with reproach, albeit with some confusion. The most noticeable reactions came from kids in my neighborhood: my association with Linda violated social and racial codes. I had enemies. "Greaser," "spic-lover," this latter as if Linda was Mexican—I heard those names. The week before Christmas vacation, a kid named Steve, whose father owned a plumbing company, knocked me to the ground. I fought back, despite the fact he was older and nearly twice my size. Black eye, sort of, definitely a split lip, visible the next day at school. Two of Linda's friends, older, one a high school dropout, found Steve on his way home from school alone and, as Linda reported to me, "beat him until he was crying like a baby."

Linda and I walked uptown, too. We spent hours after school and on Saturdays walking. We looked in storefront windows, made fun of the mannequins in Rosenberg's Department Store and of anyone on the street, particularly women affecting airs, aloof like the mannequins pointing in one direction with stiff plastic fingers while gazing to the heavens with glass eyes. We followed Fourth Street, Santa Rosa's Main Street, from the old train station at one end, just past a string of pawn shops and the Silver Dollar, a corner dive with black-painted dollar signs embossed on the saloon-style swinging front doors, back up to Rosenberg's, and further on to the Flamingo Hotel,

Iris

which marked the other end of Fourth Street with its revolving neon flamingo atop a freestanding sixty-foot tower. We rested in Old Courthouse Square, sat close together on a bench. Unsure of what to do—I was unsure of what to do—at this juncture, as if it was the next step in a ritual that had begun with the walking, we continued talking about people, passersby. This one walked like a duck, that one's socks were different colors.

"I used to think they were married," Linda said one day after school, and then laughed at herself.

I hadn't been watching. When I looked, I saw two old Indians, sixty plus, him in a sport coat and fedora hat, her in a housedress and scarf, making their way up the street. It was getting late, the streetlights already on, and the two of them seemed to have appeared from the darkness behind them, coming slowly into the light of the square. They seemed unaware of one another, or of anyone else, gazing at everything and nothing at the same time, as if they were lost, two old people in a strange city, or children abandoned at a fair. They stopped at the corner, and after the light changed and they crossed the street, they were gone, into the darkness, just as they had come.

"Yeah, looks like it," I said, not sure about what else to say.

The woman was a friend's grandmother; the old man her brother. I'd met the friend while hanging out with Linda and often visited his house, a small place behind the fairgrounds, where the old man—Uncle, they called him—sat on the front porch dressed, even on hot Santa Rosa afternoons, in the same pleated sport coat and fedora hat. He was a big man, heavyset and solid. He could do anything, the friend said—Uncle was an Indian doctor. He pulled a bird's leg bone out of his sister's eye once. He could see the future by holding hot coals in his hands. Then, one day, when a bunch of us piled into a car with an older kid, after we left Uncle in his aluminum folding chair on the front porch, there he was minutes later, two miles uptown on a bench in the square, waving to us as we stopped for a red light—I saw that. "Uncle's got wings," someone whispered when the light changed.

Mabel told me, many years later, that Uncle was the last of the old-time doctors, trained here on the earth. They followed a strict regimen of abstinence from meat and sexual relations. "Lots of rules," Mabel said. And, yes, they could do stupendous things, like traveling as fast as a hummingbird. In fact, Uncle spooked Mabel once. She was with Essie Parrish, the two of them enjoying a soda at the counter in Thrifty's, when Uncle, sport coat and fedora hat, appeared outside the window. Attempting to escape him, the two fearful ladies boarded a bus, only to find him at the other end of town, waiting for them when they got off the bus. "Playing with us," Mabel chuckled. When my friend told me the stories about Uncle, even after the incident I witnessed in the square, I said nothing. It felt like none of my business, and if I were to ask questions, it would only highlight the fact that I was an outsider. If Linda had heard similar things, still it was not my business to remark or ask questions—she was an Indian. I kept thinking about Uncle and his sister, though. What of the manner in which the two of them appeared just then, out of nowhere, walking with the measured steps of old people yet effortlessly gliding, or was I imagining as much because of what I had witnessed and the stories I had heard? Where were they off to on a cold winter's evening?

It seems funny now when I think of those junior high school days. Linda's father and my father were second cousins, something like that, both grandsons or great-grandsons of Tom Smith, the famous, sometimes infamous, Coast Miwok Indian doctor known as much for his supernatural feats, which include causing the 1906 earthquake, as for his many wives. Was he something of an "earth doctor" like Uncle? Seems he didn't abstain from sexual relations. He had over twenty children. One of his wives, Linda's great-grandmother, was Kashaya Pomo from the Haupt Ranch in northern Sonoma County; another, my great-great-grandmother, was Coast Miwok from Tomales Bay in Marin County. The trick and circumstance of history: I had a crush on Linda. One generation and the connection was broken. We were complete strangers. I was white and lost in her world. I was adopted; rumor had it that my natural father might be Mexican, a mantra I repeated to Linda. Never mind—blue eyes, fair skin, from a good part of town, I was white. Linda was Indian: her mother was Indian too, Coast Miwok, in fact the granddaughter of Maria Copa, who not only assisted Tom Smith doctoring the sick, but, with him, helped a U.C. Berkeley graduate student compile descriptions of Coast Miwok traditions and a vocabulary. Spring activities, spring traditions—the history Linda and I share which I think of now also. Tom Smith, for instance. From pictures I have seen he was a stout man, broad-faced with heavy, you might say weary, eyes; in one picture, where he is standing, it looks as if his left leg is slightly bowed. I imagine him atop his roundhouse on Jenner Point, above the wide mouth of the Russian River, facing east to the first light of day, visible above the jagged line of hills. Naming oaks, buckeyes, berries, clovers, willow, peppergrass, angelica; and animals, too, and birds. Maria Copa was there. She heard the names of things. She watched the river find itself in that light, twist and lengthen to the sea. Paths everywhere...There were lots of coming-out ceremonies then, songs for after a girl's first menses, new-woman songs, and the secret cults, the songs and arduous tasks, whereby a boy becomes a man.

I had to kiss Linda. It was expected. Enough walking, enough talking. I wanted to kiss her. I planned to make my move on a Saturday night. Did tough guys plan such things? Us kids hung out inside an abandoned garage at the end of Sixth Street, just below the newly constructed 101 freeway. It was next to Randy's house; Randy, who was white but real tough. His sister knew Hells Angels. His father tended bar at the Silver Dollar, two blocks away, and his mother, who drank at the bar, often grew impatient waiting for his father and wandered home, where she'd push up the kitchen window and holler hoarse-voiced for Randy to "come down outta there," meaning the second floor of the garage. Each time I heard her, I remembered the story of how she once struck one of Randy's friends with a hammer. We had a transistor radio up there, a table and chairs, and mattresses placed strategically in dark corners. I was even thinking about which mattress we'd end up on, which side of the room.

I went to her house first. As if, at fourteen, I had arrived to escort her to a formal dinner or something. We sat on the couch, a respectable distance apart. At one point, her mother came into the room, and then, without acknowledging me, without saying a word to either of us, returned to the kitchen, where she was playing cards with a couple of her sisters, Linda's aunts. Linda said her mother knew my father, my adopted father. I was excited: apparently, Linda had talked to her mother about me. I told Linda my parents were divorced now, my mother was working at JC Penney to help support the family. I thought that the family hard luck would impress Linda. "My father, all he did was drink," I said. Then I overheard her mother in the kitchen say something about "big eyes." Did she mean me? Was I being nosey? Is that what her mother thought of me?

It was the end of February then, maybe the beginning of March. It was cold outside. We walked in the dark, under a canopy of bushy trees that lined the street. Upstairs, in Randy's garage, it wasn't any warmer; and the two candles, upright in coffee mugs, didn't give much light either. We sat at the table with the others—Randy and his girlfriend; two Mexican guys and a Mexican girl—and talked in hushed voices. A song played on the transistor. Randy smoked. He talked about the end of the world: a friend of his mother's read a prophecy in the Bible and had determined, from certain current events, that the world would end at twelve o'clock New Year's the following year. Randy talked a lot about the end of the world up there, and then we did, too. I would be fifteen when the world ended. I heard voices in one corner, on a mattress, but didn't dare look to see, or ask, who it was—I didn't want to be nosey. It was freezing. I was holding Linda's hand. Linda finished a cigarette, rubbing it out in the ashtray with her free hand, then said she was still cold, which I took as my cue: now we must head to a mattress. Which one? I couldn't stand up and have a discussion about it. I was supposed to know. I let go of her hand and took hold of her elbow, securing her arm, without knowing where I was going to lead her—should I ask first if she wanted to go to a mattress?—when, all at once, I heard Randy's mother. I was distracted, consumed with worry over my predicament with Linda, and it wasn't until I heard "shit"s and "sons of bitch"es and saw the swirl of commotion as everyone was fleeing the table that I understood Randy's mother wasn't inside the house next door but at the bottom of the stairs. Then I heard her on the stairs.

Linda was already gone. I was alone at the table. I turned, walked two steps to the window, and jumped. Not two floors down, but onto the freeway embankment—the garage was that close. I'd seen other kids jump; still, I felt brave. I climbed onto the freeway, into the bright lights and whir of traffic. It was what, ten o'clock, and the freeway, even in a then much quieter northern California, was busy. I flew to the center divider and then, when it was clear, tore the rest of the way across and came back down the other side to the safer, better-lighted part of town.

I debated whether or not to go back. Most everyone would be scattered. Truth was, I was afraid to encounter Randy's mother. I thought of Linda. She probably went home. I told myself I missed my chance.

But I was confident I would have another opportunity. Didn't we both know that? I told myself that the next time things would go smoothly. I pictured us going to the mattress, my hand still clutching her elbow, but by then I was already heading over to Fourth Street, going home.

Saturday nights Fourth Street was crowded with flashy cars filled with teenagers who hurled insults as well as flirtations from open windows—old-style cruising. I found the wide street more congested than usual, however. Cars were stopped, moving only at a snail's pace when they did move. The sidewalks were bustling, not just with the usual Saturday-night teenagers leaning against their parked cars, but with folks of all ages, families even, mothers and fathers with kids in hand. I had crossed town earlier following College Avenue, five blocks away, and totally missed the busy scene. Old Courthouse Square was jammed. There I saw folks collected before antique cars—Model Ts, Packards—that lined the square on both sides, and understood immediately the reason for the hubbub—a car show on display the entire weekend, apparently. Never mind the cold weather and rambunctious teenagers, people wanted to see the cars.

I kept on my way, passing the square. I'd had enough for the night. I was in front of Rosenberg's when I heard the jeer. Something about "get you" from out of the racket of car radios and idling engines. I paid no attention. Then I saw a reflection in the storefront window: a face framed, as if in a square box, coming out at me. When I turned, I saw it was Steve, the guy who beat me up, hanging out of a car window.

"You're dead, punk."

I froze; in fact, the whole world froze—Steve, the car, everything as still as the mannequin in the window next to me. Only the din of engines and music above the line of cars, and perhaps it was that disembodied noise that brought me back to my senses: I thought to run.

There were other guys in the car with Steve. He could've gotten out of the car then—he was in the front passenger seat and the car was stopped in traffic—but for whatever reason he didn't. Which I realized when, a good ways down the block, I turned for the first time to look. No one was chasing me. But the car had its turn signal on. I figured Steve and his friends would come back around the block for me. The car was inching closer to the intersection. I scrambled onto a side street, around a corner, nearly careening into Uncle and his sister approaching Fourth Street. They were walking in the same aloof, haphazard manner as before, as I recall now. And Uncle was wearing his sport coat and fedora hat, his sister bundled in a heavy overcoat, perhaps an extra layer of scarves—I wasn't looking. I do know the old woman hissed with admonishment, as if I'd actually smashed into her, a reckless kid, and Uncle made grunting sounds, which I took, in the moment I heard him, as a sign of irritation also. They were in front of an alley. I cut down the dark path and dove behind a row of garbage cans.

I didn't look back or up. Tires screeched in the distance, horns. I crouched in the dirt, as comfortable as possible. Ten minutes later headlights shone at the start of the alley. The car rolled toward the garbage can and stopped. Was I a magnet for my own doom? I didn't think Steve was able to see where I went; certainly he couldn't have seen me go into the alley once I was off of Fourth Street, much less hide behind these particular garbage cans. Stay or flee? All at once, clunk, a bottle landed squarely in the can next to me. A woman's voice escaped the still open window, "Not here, James," and the car rolled away.

The next morning, arising late, I found myself outside, alone in an astonishingly warm day. The sun was daffodil yellow. The night had rolled seamlessly into this moment, it seemed. My mother had left for work, my siblings off wherever. Sunday morning. I sat on the street curb and lit a cigarette. Across the street, in the neighbor's yard, an iris grew up, rich purple. Flag lilies, people used to call them. Indeed. I went over and smelled the deep inside of the flower, fecund, then sat back down, now facing my mother's house. Right then I was certain I was a tough guy. I could take care of myself. Iris.

The night before I'd kept hidden an hour, if that. Of course, I'd like to have thought Steve found me and that I cleverly got away, maybe even fist-fought him and won, or at least rattled my bones in the cold until the first light of day. As far as I knew, Steve and his friends never came close to the alley, if they even bothered looking for me. I don't know; I never again encountered him. I walked home following empty streets.

Maybe it was a confluence of things that made me feel so confident that next morning—the weather, a harbinger of spring if not a proclamation of its arrival; the cigarette between my fingers; the iris. I ended up kissing Linda once, that is, before she took up with

Iris

an older, much tougher guy than me. She's gone on with her life now. I'm sitting here remembering all of this. As it turned out, I would get to know Uncle and his sister better; certainly his sister, who in her latter years, before she died at the age of 101, I came to call Grandma. The family tells me that Uncle and Grandma used to feel sorry for me, that I was always one of their favorites. Hobo boy, they called me. That night, at the entrance to the alley, I didn't think to imagine they as much as recognized me even. But, if the family is right, might Uncle, in his sport coat and fedora hat that night, not have been grunting in irritation but instead singing a lost boy on his way?

STEPHEN MEADOWS

In the Mountains

In the cold March rain
ponds crater and roil
Time passes
I can manage no poem
though I've filled many pages
with notes about sorrel
wood violets
and the sound of creek bottoms
hissing the name of this storm
The sycamore flings its abandon
over the strict wet road
Alder and willow too
tiny buds firing
into evening
a myriad of quail

DEBORAH MIRANDA

Water

gets what it wants. Water finds
a way. It seeks out the possible

crack, the promising path
of least resistance. Water refuses

to believe in walls, stone, root,
isn't afraid of flood-force.

Water makes a way:
pushes over, under,

around, through. Water rams
sand and rocks down

the ribbed black throat
of a culvert,

busts through terracotta stones
and thick mortar, tosses

rubble like redbud petals, swarms
over the broken remnants of restraint

and rushes the road like a glorious army,
exulting with gravelly April song. Water

gets what it wants, one way or another.
Water dreams of storms

like you.

DARRYL WILSON

Kweme Psukitok, Spring Maiden; *Amal*, Flower Maiden

Many wonderful powers came and made our world. *Kweme Psukitok*, Morning Girl, *Amal*, and Butterfly Girl all came to make beauty and wonder and magic. This is a story about *Kweme Psukitok* (Spring Maiden) and *Amal* (Flower Maiden) so you will forever have a greater appreciation for the magic of life all around us when flowers make the world beautiful for children and hummingbirds, bees dream and make honey, and blossoms prepare to magically turn to fruit.

Do not think that Spring Maiden and *Amal*, Flower Maiden, are the same power, for each has a separate duty and they each have delicate yet different destinies.

Long ago it came to be that robin had a dream. Dream told robin how to construct a special nest for a special singing that would still be a safe place for her children to grow, so robin complied. Spider had a dream—to wrap many strands of magic twine, tying nest to tree limbs—so nest would not fall to earth. So it was that summer passed, autumn appeared and passed. Little robins flew away. The leaves fell from the great oak tree, revealing little nest to winter winds. Wind blew and whistled, causing little nest to be loosened from limb. Wind gusted and whipped. But for the spider's silken strands, nest would have fallen to earth. Nest hung there between greater limbs, much like a hammock hangs between two strong trees.

Winds continued and nest was there swinging back and forth. A covey of children scurried by. They stopped and talked with each other and with nest swinging in the breeze. After a short while they hurried on, throwing passing words to little nest. Nest fluttered back to the children.

They say that is as close as we have been to seeing Spring Maiden and Flower Maiden. The children said *Kweme Psukitok* and *Amal* were seated cross-legged in that nest swinging in the winter winds, that they could be seen and at the same time could not be seen, and that they sang beautiful songs to earth. It is said that at this time they planned for life and for making flowers to make earth pretty. They sang sweet songs to the seeds still folded in winter sleep. Whistling wind carried their melody everywhere.

Amal sang songs so sweet and beautiful that song reached into the heart of seeds yet wrapped in winter's blanket, giving each seed a more brilliant flower to dream of and a greater destiny. The song gave to each little lump on the limb that turns into a blossom a sweeter flavor and a deeper responsibility.

It is said that it was a wonderful song, long ago, that caused the creation of our world. Still today song makes beauty and wonder and we can see that beauty and wonder every spring; it is there also in the fruit of autumn. The beautiful powers took turns singing there in the winds of winter. Theirs were sweet songs, making the flowers want to hurry and the blossoms want to bear fruit.

Ju'wa, grandmother, rested. A child asked a question.

"*Ju'wa*, if the wonderful powers of spring and flowers are always singing, can we hear them?"

"Child, in the chill before dawn, when stars are brightest and eastern horizon silver-green, go out into the hush and listen. Songs will come like a flute melody that you can hear only with your heart. Listen with your entire being but mostly with your heart and you will hear the singing."

So we went out before dawn as instructed and faced the vast universe, listening intently. Yes, the songs came, dancing through the stars, on an ancient flute melody. We cried and as our hearts listened, we thanked *Ju'wa* for her precious words of much wisdom. Because it was beautiful, we cried again.

SUMMER

YOKUTS

Rattlesnake Ceremony Song

The King Snake said to the Rattlesnake:
Do not touch me!
You can do nothing with me
Lying with your belly full,
Rattlesnake of the rock pile,
Do not touch me!
There is nothing you can do,
You Rattlesnake with your belly full,
Lying where the ground-squirrel holes are thick.
Do not touch me!
What can you do to me?
Rattlesnake in the tree clump,
Stretched in the shade,
You can do nothing;
Do not touch me!
Rattlesnake of the plains,
You whose white eye
The sun shines on,
Do not touch me!

FRANCISCO PATENCIO

from "The Quail Legend"

Long ago in summer when it was dusty and dry, an old blind man was left at the water hole. He could no longer go by himself. Maybe they could no longer help an old blind man. I do not know. So, they brought him to the water hole and left him there.

When the Quail came to the water hole they were talking among themselves as always. The old man thought they were people and spoke to them.

When the Quail saw the old man and saw he was blind they said, "Here is an old man standing. Let's take him with us."

So, they all left the waterhole and started walking to the east. As they went they sang their song to the trees, the Greasewood, the Cat's Claw, the Desert Willow, the two kinds of Palo Verde. They led him to a spring and said to him, "Here is good water."

When it was getting evening and darkness was coming on, they told him that it was time to lie down and rest.

When the daylight came they were up and said to the old man, "Daylight has now come. We must be on our way." They then started to walk still to the east.

When the sun was high in the sky, they said to the old man, "It is the middle of the day. It is very warm now."

Then the dust and the wind came, rattling the sticks and the dry weeds. The Quail could have flown away, but they walked slowly because of the old man.

Finally they came to Yuma. They led the old man to a house and said to him, "You are now standing by a door. Knock on the door." Then they went away.

from "The Quail Legend"

    The old man had learned the song of the Quail while he was with them—the song of where they went, where they lay at night, when they rested, and what they found to eat. He taught this song he had learned to the Yumas. They learned it and liked it very much. Then other tribes learned it too. Soon it became known to many more tribes. They all sang it in their own language.

DOROTHY RAMON

Why People Did Not Kill Tarantulas

'Aam tecqwam 'ip kiikam waha' wuuwerham.
There are a lot of tarantulas living here.

'Aam waya'xq pavay'pa' te-er'cu'ow.
They come out after thunder showers (in the summer).

'Apya'vu' hiit te-er'c, 'apya'm wangatk peekinu'.
After it thunders, they come out of their holes.

Weerr, Weerr nemey 'ip tengek. Tecqwam keym 'aam.
A great many of them walk around here. They call them *tecqwam* 'tarantulas'.

Xhinyim kukam per'pernavem. 'Im tewva' qac: tewva' kiikam peerherv.
They are like spiders. They live underground in holes.

Peeki' tewva' qac.
Their home is underground.

'Aam kwenemu' 'uviht Taaqtam qaymu' qeern 'aam maqayeewp kwenemu' taaqtam teer hiit 'aqwahaynivey payika' tum hayp mermerher' yu'pa'.
Long ago the Indians did not kill them because they would tell people when plants were ripe in different areas out in the wild.

Why People Did Not Kill Tarantulas

Puyaan peerraqwi' teer 'ap kwenevu'.
He (the spider) would tell them about their food sources far away.

Peerraqwvu' qwahow, key kwenemu' pemeka'. 'Ayee' taaqtam 'enan 'amay.
They would tell them when their food was ripe. And then the people would know.

'Ayee' myaawnkim chaw'cu' 'amay hiit hami' peerraqwi' qwahow 'ap hayp 'ip qaqaayv tum hayp.
And then they would go pick their food when it was ripe up there in the hills.

Pana' kwan nyii'ac 'ama' tecqwac.
That's what the tarantula used to do.

'Amatunga' qaymu' haym qeern 'aam tecqwam 'uviht.
That is why no one killed those tarantulas long ago.

Qaymu' haym haypa'n qeern 'aam tecqwam. Wuuwerham chevêkim qace'.
No one ever killed those tarantulas. There are a whole lot of them.

Wuuwerham tum hayp mermerher'.
(There are) many, all over the place.

'Aam 'aame' tum hiitim teer pana' nyaawnk maqayeewp tervav kiikam.
They (the tarantulas) were able to tell them because they live under the ground.

Tervacvu' teer hamya'qac hiit hayp qwahow hiit hami'. 'Ama' 'ayee'.
The earth would tell them where the plants were ripe. That's all.

DEBORAH MIRANDA

In Praise of August

She is a jar of jam just filled
right up to the rim,
a dense burgundy pulp,
picked tart off bristling red vines.

Dip up a spoonful. She bursts
across your tongue with droplets
of thunder, lightning, burning rain.

She rests on the counter, heavy,
steaming into air already heated
beyond the last degree.

She just might shatter into shards
of scalding joy.

GEORGIANA VALOYCE-SANCHEZ

Summer 1945

Summer   1945
rural East L. A.
the hills   a gathering
golden quail   waiting
uncut fields of hay swaying
in the wind   around
my Japanese schoolhouse
home
         abandoned   they said
the wood-frame schoolhouse
mostly one large room with
open wide windows beneath
a gently curving roof
         simple lines
a Japanese painting in
the morning mist

That summer
we lived in one side of
the schoolhouse
         lazy days   playing
on the long front porch
shaded by pepper trees
and I was too young to
question   where
all the children
had gone

SUMMER

At night
coyotes howled
and far   far away
bombs fell on Japan and
the whole world was screaming
bodies gutted   mutilated
arms and legs and heads
torn   blown apart
survivors wishing they had
died   not knowing
that soon
a new bomb would fall
and they would lift their faces
to relatives   falling
in black rain
   and I slept peacefully
safe   under a warm quilt
beside my sisters
   the hum of our parents' voices
as warm as the wood stove
Mama cooked on
And the day
the bleating wave of sheep
swept   down the hill
like a flash flood
rumbling towards
the schoolhouse
no matter   the black dog
   barking at them
no matter   the old man
   in the dusty brown hat
   and stick
   poking them
no matter   the shouts
   and running to swerve them
   aside

Summer 1945

they came
    pouring through the open windows
    through the doorway
    bumping and trampling
    each other in the playroom
    milling and crying
    trying to escape their sudden
    confinement
and the Japanese schoolhouse
shook   with their awful
confusion
the wood creaking
and moaning
like a tortured accompaniment to
a Kabuki tragedy
    and I was afraid
in my safe side of the house
    even as they spilled
over the windows and porch
and out onto the dirt road and
they were only   a dust cloud
rising
in the distance

STEPHEN MEADOWS

In the Water over Stones

for Isabel Meadows

Your voice Isabel
is a quail's voice
as the sun's song ticks
in the brush

It is the hawk's voice
and the heart's heat
of the rabbit
in the parched summer grass

Nearby in the river
in the water over stones
it is a willow voice
it is a crayfish voice
in the hollows
in the darkening places

At first light
it is the wind's voice
the mouth of the river
tule voice the voice
of a hundred breezes

the sun marks out
the red madrone

In the Water over Stones

and in the canyons
it is a redwood voice
a sycamore voice
sweet scented

In the spring
it is the lupine voice
a blue white and purple
coverlet voice
all over the hills
and the meadows

On the river banks
as the set fires burn
and the steelhead run
it is the hunter's voice
flinging the gleamers
silver on the sand

Though the houses
of rich men now cover these hills
it is your spirit voice
your evening voice
your voice of the western waters

The stars hang out
over the point of wolves
on the edge of the world
the sea lions call
the otters break open
abalone

It is the voice of the land
It is the voice of bright shells
It is the voice of the valley
And the mountain Isabel

It is the voice of the people too

It is the weaver's voice
It is the young girl's voice
The gatherer's and the singer's
and the farmer's voice
the wives and the children
and the old woman's voice
It is the Indian voice
and the whalerman's voice
and the voice of the servant
escaping

It is the voice of your face
across the years Isabel
in my grandfather's face
in my father's face
and in my face as well

It is the voice
of the ones on the edges Isabel
It is the voice
of those ones with no voices

Hawk and rabbit
Quail and brush
Water and willow and crayfish and stone
Wind in the canyons
Daylight through limbs
The lupine the steelhead
The cookfire's call
Beans and tortillas
Your memories Isabel talking
Talking to us all

SHAUNNA OTEKA MCCOVEY

Fireworks

I didn't imagine I'd see you
when I did, (I mean *see* you)
at the foot of the stairs, you turned
and smiled, your eyes sparked

like fireworks on the 4th of July.
I'm not even patriotic, yet I
compare them to that brilliance,
I felt the freedom so coveted, the

independence, the liberation
sought by many, practiced by few.
What have you done? I am
speaking in red, white and blue

the blaze of your smile ignited
a soul thought dormant.
I'll soon be baking apple pie
and waving stars and stripes

dreaming of big, beautiful
explosions of you—I exploded
when I *saw* you that night,

this new American/Indian
with hand over heart pledges
allegiance to you.

ROBERTA CORDERO

Haiku/Senryu

4th of July

Independence day
occupation made holy
glares red and bloody

Water-Strider

Water-strider walks
surface tension undisturbed
dimples in its wake

Heron

Herons blue and green
feathered summer spirit guides
gracious visitors

GREG SARRIS

Osprey

Summer.
Something about the glare of noon. Or nearabouts noon because summertime that hour feels like eternity, the essence of the season itself, halfway between here and there, stopped. Something about the stillness of light, and the motionless surface of the green river. On a dry path above the water even the orange-flowering monkey plant and sticky-leafed mountain balm appear to be waiting, as if for the sun to move again. Until an osprey breaks out of the sky, silver body and black-tipped wings, coursing the snaking path of the river.

All at once, it's the same light, the season again forty years ago, and the girl below the Ferris wheel at the county fair will come along the path with the boy just now heading for the carnival from the livestock barn. She does not know him; wouldn't imagine ever knowing him. But the light is open, empty all around her—nevermind the throngs of people, the cotton candy and hot dogs, or the shrieks pouring from the two-story Ferris wheel behind her—and anything is possible. Her friends have gone to toss rings for a stuffed bear. She is alone. She is free.

She sits on a bench and pulls a compact from her purse. She wears a pink and black polka-dot blouse with a ruffled front and a black fitted skirt she borrowed from her older sister. She is Indian but she could be Annette Funicello. She could be different. Not like overweight Lynette, whose legs are like stovepipes, or Betty, her other friend, with cat-eye horn-rimmed glasses, poor thing.

She would say it was all like a movie, how, when she looked up, after snapping her purse shut, he was there—her lipstick was on and she looked perfect—and he walked up to her, and he had a car.

She knew better, she would say, too. She wasn't a fool. She was sixteen, after all. She just wasn't thinking. He talked while he drove. He was from Petaluma, on the outskirts. His family had a dairy; he showed cows at the fair, registered Holsteins. That morning, first show, he had won a blue ribbon. It was his lucky day, he said. He'd met Mexican girls before. He liked them. She didn't tell him her name, that it was Linda, and that her last name was as American as all get out, but if he had asked and she had told him he would know she wasn't Mexican, or certainly not all Mexican.

He called the place Wohler Bridge.

She knew it only as the bridge on Eastside Road, the place where Filipinos once held cockfights, big get-togethers, where Indian women went to meet those Filipino dandies in pin-striped suits and Panama hats, handling bets and squawking, razor-fitted roosters with equal aplomb. Which she thought of after the boy parked the car and they traversed the dusty parking lot, heading for the trail—how last summer, or the summer before last, her aunts pointed to the place and mentioned as much while driving over the bridge on their way to pick pears someplace.

The boy led the way. Coming around a bend, she saw the monkey plant and mountain balm alongside the path, orange flowers, clumps of sticky leaves. Nothing moved there. Up ahead the path narrowed into a dark copse of willow. She looked back and could no longer see the parking lot or the bridge spanning the river. Then she remembered what her aunts had said, how, finished with their stories about cockfights, they cast their gaze in this direction and were quiet, if only for a moment, before they told the story.

The boy must have thought she was crazy. Without a word, she turned and walked in the opposite direction, as if somehow she could get back to the parking lot without him noticing. There, in view of the bridge and the well-traveled road that led to it, they talked, never mentioned her abrupt about-face. They skirted talk of fear or dashed hope. He was a gentleman, she said, this handsome white boy with a mess of close-cropped blond curls and powder blue eyes. His small mouth widened unbelievably, unnaturally even, into a broad white smile, which he wasn't doing so much now.

"I didn't know what to tell him," Linda told me, "how to describe what I saw."

She paused, shrugged her shoulders. She asked if I wanted more lemonade. Already, at ten in the morning, the temperature outside was ninety degrees. The kitchen, where we sat, was stifling, even with the windows pushed open. The apartment was on the second floor, which didn't help matters, and above a corner deli: I smelled onions and fried meat. Every now and then, our conversation was punctuated by the ring of an old-fashioned cash register.

"What do you mean?" I asked.

"The whole story."

"What story?"

Then she said what she saw.

A woman on the path. Coming with the man wearing the straw sombrero. Her hair, which was fixed for Sunday mass at St. Rose Church, though jet black, resembles Rita Hayworth's flowing red mane in *Gilda*. She wears a plain skirt and blouse set off by a wide patent leather belt and matching shoes. Her stomach swells slightly below the belt; she's not pregnant, thank God: today or tomorrow she will menstruate. Thank God, because she has six children, hard enough already.

She was both surprised and thankful that the man wearing the sombrero showed up at her door last night. Surprised because she had wronged him, exercised poor judgment with a co-worker in the hop fields, poor judgment he was made privy to; thankful because how many men would be willing to care for a woman with six kids, and then give the woman a second chance.

That was what he had said in the dim porch light: a second chance. He'd come directly to her house after twelve hours mending apple crates in Sebastopol. Their lovemaking was rigorous. In the morning they went to church. He had only his work clothes but she had ironed them. He showered and used her oldest boy's aftershave. He left his hat on the pew when they took Communion. After, he wanted to go for a ride, just the two of them, the kids could take care of themselves.

She knew a place.

She picked mountain balm with her grandmother and great-aunts there. She remembered summer days, and the sharp scent of the herb, but, most of all, the appeal of the slow-moving green water. She knew about the cockfights—what Indian woman hereabouts didn't?—which was how she knew where to direct him off the road to park.

Just north, upriver from where they left the car, there was a wide, empty circle of tramped earth. A column of fennel, like a gatepost, grew alongside the trail. Chicken feathers specked the ground and clung to clumps of chaparral that loosely enclosed the circle; gold and red and black feathers, like leftover decorations. A faint line of smoke wafted from a heap of ashes. The still emptiness of the scene, felt in the noonday light, emboldened her. Crossing the circle, she felt alive, complete, as if she had been present at the cockfight the night before and alone survived its chaos. She was not carrying another man's baby. They could start over, a clean slate, her and the man wearing the straw sombrero.

They pass the orange flowers and taller, sticky-leafed plants; no doubt, the man is leading now, because he sees, up ahead, where it is they are going. She isn't paying attention. Maybe she's glancing down at her shoes, seeing how dust has filled the cracks in her patent leather: minuscule etchings, like rivers drawn on a map. And, all at once, in the blink of an eye, they are in the willows, the woman and the man, and she is looking back to the light.

Did she feel a chill in the shadows, and remember again her grandmother and the herbs, not the sun and warm days but wintertime, when she had a cold, felt a tightening in her chest, and the old woman was taking the dried leaves from a Mason jar to boil tea? Or maybe she was just looking back up the trail, if only for a way back to the light? Or maybe she didn't have time, even, to comprehend what he was saying, that no other man would have her, for it happened so fast, one of his arms securing her head, the other forcing the hot blade into her swelled abdomen again and again.

Linda knew details. Did her aunts read the man's confession in a newspaper? Did they see a photo of him? Or had an ordinary and horrible story grown legs and feet?

"I didn't really see her until I was back at the car," Linda said. "It was just this thing that came over me…Stupid, I guess. That boy was cute. He must've thought I was an idiot. We kept talking—How could I tell him I was seeing that woman looking back at me?"

"Maybe she thought of her moon," I suggested.

I figured she wouldn't engage, at least not favorably, the moon idea—that the tragic woman, in her last moments, thought of the menstrual taboo, that not only during her menstruation, but also immediately before and after, a woman mustn't hike in the brush, particularly near a body of water, such as a river—but I wanted to impress her with my knowledge of such things. Linda was proudly Catholic; though aware of Indian traditions, she mostly scoffed at them as backwards.

We were cousins, both of us descendants of Tom Smith, the legendary Coast Miwok medicine man, though at the time neither of us knew. More than anything else, we were friends: we'd met almost ten years before in junior high, when I had a crush on her (an older suitor eclipsed any chance I might have had). I'd moved to Los Angeles a year and a half earlier and was home in Santa Rosa visiting. She had acquired the apartment recently and wanted me to see it.

My mind kept whirling with the story she told, imagining the woman's state of mind, the significance of the gruesome tale—and Linda's experience too, what it meant for her at the time as a sixteen-year-old. She didn't preoccupy herself with so-called Indian lore then either, but hadn't certain notions crossed her mind? Even now, seven years later, wasn't she repressing thoughts of taboo?

As if she was reading my mind, she answered dryly, "What difference does it make, a woman got murdered, isn't that enough?"

I was gazing out her window, to the rooftops across the street. I looked back to her, if only not to be rude.

The first time I had been to Wohler Bridge, passing over it, was with Mabel McKay, the renowned Pomo basketmaker and doctor, who taught me about everything worthwhile I know. Again, it was summertime, the year before I moved to Los Angeles. In those days, I often drove Mabel to pick herbs or dig sedge for basketmaking. If I remember correctly, we were on our way to Dry Creek to explore the sedge beds (now below Warm Springs Dam's three hundred feet of water).

As always, she was noting features of the landscape. She nodded upriver and said that she had heard of a good place to pick mountain balm there. Was it her longtime friend Essie Parrish who told her of the place? We were somewhere in border territory, historically shared by Coast Miwok, Kashaya Pomo, and Southern Pomo—obviously not in Mabel's native Cache Creek Pomo region of eastern Lake County. But what I remember is her mentioning how, at that time of year, the mountain balm, indeed many herbs, would be mature; "mature time" was how she put it, and I was struck, finding not only her use of the word but her tone more suitable to a group of well-behaved schoolchildren, or perhaps an elder, say a woman after menopause.

When I visited Linda, I was on my way to Wohler Bridge, on my way back, I should say, for a second visit. On a tip from a friend I'd learned of the place as a haven for nude sunbathing. It was 1974. I was twenty-two years old. I didn't tell her that I had been to the place, or that I was on my way back, only that I'd heard that young people were hanging out there, nude sunbathing, which was what prompted her stories. We had gone our separate ways, her to a job, me away to college: I wasn't certain how she felt about the idea of public nudity and didn't want to offend her sense of propriety.

 That seems so long ago. Recently, Linda let me know, with an ironic chuckle, that she was aware of my visits to Wohler Bridge. A mutual friend described my presence there, Linda said, "in a flattering manner." Now I was the one embarrassed, a middle-aged man face-to-face with a middle-aged female cousin and childhood friend. Today we know our connection to Tom Smith, how exactly we are cousins. And I know many stories that connect me to the place in specific ways: Tom Smith and other Coast Miwok ancestors followed the Russian River inland from the coast to pick herbs there (long before the appearance of a bridge), mountain balm as well as angelica, which once grew in abundance in a damp recess nearby; and, many years later, my Filipino grandfather, with my Coast Miwok grandmother (Tom Smith's granddaughter) on his arm, watched the notorious cockfights, saw feathers fly there.

 I returned to Wohler Bridge several times. I watched the crowds dwindle, particularly as the door closed on the now unbelievable era between the pill and AIDS, and then altogether disappear when the property owner, reputed to have been the late actor Fred MacMurray, fenced off all access and posted very noticeable No Trespassing signs.

Osprey

Older now, I am nostalgic for those days of bliss in the sun. I went there to relax, escape worries of school and whatever else, to forget.

This morning, thirty-five years since my first visit, I returned once more. Driving along Eastside Road, across the bridge, adjacent the stand of redwoods, I discovered a parking lot, then, past an open chainlink gate, a trail leading back to the river. An Asian couple stood fishing on the shore, dungarees rolled up to their knees, crude bamboo poles extended over the green water. Farther upriver, parked against a massive redwood trunk, one man, and then not far away, another, both clad in flannel shirts on this hot day, stared out at the river and beach across. Were they waiting for fun reminiscent of the old days, or just remembering it?

I forged the river, swam across. Still a beach; no people on this side but still a beach, sand, the water. And of course that eternal summer light. I plopped down, felt the beach sand, cool water dripping down my shoulders.

The osprey, that magnificent bird that burst from the empty sky, broke my bliss. I watch now as it follows the river and vanishes as effortlessly as it appeared. I've seen ospreys here before. Like this one, they always seem to come from nowhere. A knowledgeable fellow sunbather from days gone by speculated that ospreys had a nest in the redwoods across the water. She said they had great vision, able to see far distances and into the water many, many feet—their eyes are shielded from the sun's reflection on the water's surface. They can live thirty years or more. Had I seen this same bird before? Was it now searching the water for a plump carp?

Still, even in the osprey's absence, the place stirs; the world is set in motion again and the hours march slowly on. The faintest trace of shadow. The monkey plant and the mountain balm, leaning, anticipate the afternoon. My mind is wild with stories.

I think of them all: Linda and the farm boy with the blond curls; the woman with Rita Hayworth hair and the man wearing the straw sombrero; my grandfather and grandmother and the swirling dust of fighting roosters; old man Tom Smith, one leg slightly bowed, picking mountain balm leaves. Then, just as fast, I remember Mabel, not when we drove over the bridge, when she said "mature," but her warning me—I hear her actually—saying not to tell stories after and before the frost, after and before winter season, not only because you must pay attention to where you are going, watchful for snakes and such, but because you too are coming out, becoming story. It was the rule, she said, and added, for my benefit it seemed, that it was silly to think of stories all the time.

But the stories won't let up. They converge in a moment. My own story, I think, what happened when I came back here after visiting Linda.

I see it.

Coming along the dusty path, the orange flowers and sticky leaves coming into view, even the copse of willow up ahead, and I'm thinking of the stories I have just heard. But, no matter, because I don't follow the path; turn to the beach instead.

There are lots of people: unclad bathers lining the sand.

It is a glorious day, and I am glorious in it. I am twenty-two. Earlier, before my stop at Linda's apartment, I paid a visit to the gym: I still feel pumped up. My seventies stylishly long hair, mustache…But I wasn't thinking about stories, and this is the same story again. No, not the set of eyes I felt upon me, even when I turned and glimpsed beside a cottonwood the spectacular form clad only in a string of beads, telling myself to compliment the beads, talk about the damn multicolor beads, not even the line drawn in the sand by the shade. But then I stopped, before a single thought. Then, for however long, that moment, I was more alive and smart in this place, indeed in the whole world, than ever before.

That was summer, I am thinking. Knowing summer for the first time.

Then my reverie is broken by the sound of a truck approaching on a dirt road behind me, and I remember that I am trespassing. I dive into the water, heading for the other side.

FALL

ISHI, *translated by* LEANNE HINTON

from *Ishi's Tale of Lizard*

At dawn,
Lizard took his quiver and his storage baskets.
Now he went west,
Went west across the water,
Went west up the mountain.
He put his quiver down on the ground.
He climbed up to get pine nuts.
He climbed back down.
And then he piled pine-nut cones all around the fire.
Now he worked at pounding,
He pounded the pinecones for nuts,
   That's what he did.
He picked up his storage basket,
Picked up still another basket,
And now he gathered up the nuts.
He took them up in his hands.

A great sound descended.
The Yawi shouted their war whoops.
"Ho, I think
It is the wind," he said,
Pretending to ignore them.
"It looks like rain now.
I'd better sit here and shell nuts," he said.
He picked up his storage basket.
"I see you on the ground everywhere!" he said to the nuts.

FALL

He slung his basket over his shoulder
And carried it under his arm.
He took his bow.
Now he went along the trail.

LUCY THOMPSON

The Acorn

Many years ago several families were out camping in the fall, in the last part of October or November, gathering acorns for food. (When the families get all fixed up in their acorn camps, all go forth to pick the acorns each day as they drop from the tree, using the large baskets to put them in and carry to camp. In the evening when all have gathered at the camphouse and the evening meal is over, all the family—men, women and children—take their places and commence taking the hulls off so as to get the meat or kernel out. This is done by the teeth, and it is wonderful how expert we become at it; and it is seldom a kernel is mashed or bruised. These kernels are nearly always in halves, sometimes in three pieces, and once in a great while there will be four pieces; and to find one that is divided into four pieces just as it grew in the shell is not a common occurrence. There is on the inside of the outer shell a very thin skin that covers the kernel, or meat of the acorn.)

    There was a young Indian girl out with her basket picking acorns, and as she went along with her basket picking up acorns she would, as often as she could, place some in her mouth and crack the hull and take the kernel out and put it in the basket with the ones that were not hulled. As she was going along, she happened to open one where the kernel was in four parts, which at once became very amusing to her. So she set her basket down, and on taking a look at it she took the outer hull off and made a neat little cradle out of it; then she took the inner skin part and made a nice set of baby clothes. After she did this she took the whole of the kernel and covered it with the clothes and placed it in the cradle that she had made of the hull. After all was finished she looked at it and then put it in the hollow of an oak tree and went on picking her acorns until time to go back to the camphouse. When it came time for them all to return to their homes, she had forgotten what she had done. One day while she was preparing some acorn flour she heard a noise behind her, some one saying "Mother, mother," and on looking behind her she beheld a little boy; and as soon as she saw him she knew that he was formed from the acorn that she had fixed and left in the hollow oak tree. She raised the Sa-quan, or pestle, in her hand and tried to catch the boy, but he ran from her, and she followed after him; and the race kept up until the boy got to the edge of the ocean, where there was a man in a boat. So the boy jumped into the boat. The man pushed the boat off and together they started out to sea, and had got well out when the girl arrived at the sea shore. She hurled the stone pestle at them and it fell into the sea, and the top of it stuck up and is there to this day.

SHASTA

Coyote and Raccoon

Coyote was a constant hunter, but never successful. At sunset Raccoon would bring home plenty of fish and quail. Coyote came to his house and asked him, "How do you catch these fish?"

"Oh, I just hunt about in deep water. As soon as I see the water is full of fish, I go down into it with some long sticks on which I string the fish. When one stick is full, I bring it out and go down and fill another. That is the way I catch fish."

"How do you catch these quail?"

"I go about in the brush, and when the quail fly up into a tree, I set fire to the tree and lie down under it with my eyes shut. The quail keep falling down into my hand. That is the way I catch quail."

That night Coyote could not sleep for impatience. At dawn he went to the creek and searched for a deep pool. Finding one full of fish, he cut five sticks and jumped into the water. He groped about, but the fish slipped through his fingers. When he was nearly drowned, he came out with just one fish, old and ready to die. "I think that is not the right way," he said. "Perhaps Raccoon deceived me."

He went into the brush, drove a bevy of quail into a tree, set fire to the tree, and lay down under it with his eyes shut. The burning leaves and bark fell on his breast, but for a while he endured it. After a time one old charred quail fell down. "That is not the right way," he said. "I think he deceived me." He went home, and to his five sons he said: "I do not know how I shall feed you. I think I shall watch Raccoon when he goes hunting."

All night he watched the door of his neighbor's house. At dawn Raccoon came forth and Coyote crept along behind him. As the sun was rising, Raccoon went into a thick clump of bushes, and Coyote saw him taking quail out of a string snare. He tied their feet together and hung them in a tree. Raccoon then went to a place where the water was rushing down in a riffle. Coyote followed him and saw him open a trap and take out many fish. While he was bending over to string them, Coyote leaped upon him from behind and killed him. He placed the body among the fish, covered the bundle with leaves, and carried it home.

Raccoon had warned his seven sons that if ever he failed to return home at the usual time they would know that Coyote had killed him. The Raccoon boys were playing with the Coyote boys, and one of them said: "It is time our father was coming. I wonder what is the matter?" Soon they saw Coyote come with a heavy load of fish. He opened the door quickly and pushed his pack inside. But the eldest Raccoon, looking at the end of the pack, saw the face of his dead father. He told his brothers.

The next morning Coyote went away to the river, and the eldest Raccoon said to his brothers, "Let us kill the young Coyotes." So they killed them, and built a fire and roasted them, and left them lying in a circle inside Coyote's house. When Coyote came home and saw the roasted bodies in his house, he thought his sons had killed the Raccoons, and he said to himself: "That is good, my boys! You understand. You know how, my boys." He sat down and ate them. When he came to the smallest and last, he looked at his face. Some of the hair had been singed off, but he perceived that it was his own son. His heart leaped up to his mouth, and he feared it was going to pop out. He patted his breast and said, "Wait, wait, my heart! Do not come out! Do not come out!" From every corner of the house and all about outside where his children used to play he heard their voices calling. He stood up and called out: "Raccoon boys, come back! Do not run away! Your father is coming home! He caught so many fish he could not carry them!" He ran about here and there, looking for the young Raccoons. But they were climbing to the top of a tall tree. He ran to the edge of the water, and seeing the reflection of the Raccoons he jumped in after them. He scrambled out, looked down, and saw the reflection again. Once more he jumped in. Then he happened to look up and saw them in the tree. He secured a long stick with a hook at the tip, and with this he reached up, saying, "Come down, Raccoons! Come down, Raccoons!" But the eldest said to his brothers: "Do not look down! No matter what he says, do not look down!" Nevertheless the youngest looked back as they climbed higher, and immediately he fell into Coyote's open mouth and was devoured.

Coyote pursuing the climbing Raccoons is now seen in the autumn and winter sky, in the form of a large star that follows the Pleiades across the sky.

WINTU

Puimeminmak, the Deer Maker

Puimeminmak lived with his grandmother. He lived with her a long time. He grew big enough to hunt deer. Every day he killed deer; he never failed. They had much deer meat. They saved every part of the deer. Every part was hung around the house. Every part was hung in a separate place. He did this for a long time. Then the man said, "I am going to leave you." He said, "I am going south. When I have gone, there will be no deer in this country; all the animals are going to follow me, all the birds, the snakes, everything. You use up all our meat. A big snow will come. Everyone will starve. When you have eaten all the meat, go down to where the trail crosses the creek and set a snare. The next morning you will find a deer in it. That will be for you to use."

Then he left. The first day he traveled all day and then he camped that night. He built a fire, smoked, and talked to the animals he had with him. He had every kind of animal with him. The next day he traveled again. The animals hid during the day. The same thing happened every day and every night. He traveled, for about ten days. One evening he came to a place and said, "Here we are now." He said, "You animals must hide yourselves until I tell you to come out." Then he went down into the valley. It was sundown. He went south down into the valley where there was an earth lodge. He sat down. Then he saw two women, two sisters, come toward him with carrying baskets. He turned himself into an ugly, dirty little man with a bow and arrow which were of no use. The younger sister came and saw him sitting on some manzanita wood. She smelled him, jumped back and said, "Someone is here." The elder sister spoke to him. He had two mountain quail. The elder sister said, "Where did you get them?" He answered, "Oh, I just got them." Then they said they were going home. The elder sister put him in her basket. He said, "Don't tell anyone." They took him home. It was dark when they got there. They hid him so the mother and father did not know he was there. He stayed there all the next day. The elder sister gave her mother the quail. They ate it but the mother asked no questions.

The next night Puimeminmak told the two girls to sleep soundly. After midnight he got up and went to his uncle who lived quite a long way from there. He got there the same night. His uncle said, "Hello, nephew. I heard about your traveling." Then Puimeminmak asked his uncle for something to wear. Then he took a dagger and peeled off his little dirty skin and turned into a big fine-looking man with hair down to the ground. He put on moccasins, an otter skin, and beads. He stuck a dagger through his topknot. He had a bow and arrows. Then he left his uncle. He returned to the earth lodge. He had been gone just a short time. He lay down with the two women and one of them woke up. She did not see the ugly, dirty little man. She thought that he had killed him. He said, "No, I am the same man."

The people who lived there hunted all the time, but they never killed a deer. They killed snakes and called them deer. The chief made a speech and told all the people to go hunting; he asked Puimeminmak to go too. He said, "I have never hunted much, but I'll go."

They went. He looked for deer tracks but saw none. No one killed anything because Puimeminmak wished it. They came home that night and said they had killed nothing. They kept on doing this for a long time.

One day they went hunting again. They went way off in the mountains. They circled around to drive the deer. It was midday and they had not yet seen deer. Then they started home. Puimeminmak said to his father-in-law, said to the chief, "You and your men had better stay with me." Then they came to a gulch. He said, "Stop, there are deer." He shot at them and killed twenty-five right in one spot. They were real deer. The others were surprised. Each person had a deer. That night they had a big feast. Puimeminmak said, "Don't eat snakes, they are no good. You want to kill deer."

The next day he told his animals to come out and show themselves. Half of them he told to stay hidden. He was going home soon. The other people went hunting for real deer now.

Puimeminmak stayed there a long time. He had a son. Then he decided to go home. He went home with half of his animals. He went back with them. He returned to his grandmother with his wives and his son. When he returned, there were plenty of deer again. That is why every fall the deer go south and return in the new year.

DEBORAH MIRANDA

Faith

Yesterday I saw
wild turkey running
in the road,

fluttering, frantic
for cover; a young deer
dead in the stream.

This morning chainsaws whine
on storm-dropped oaks,
woodsmoke tendrils cold air.

You put the garden to bed
with bales of sweet hay
from a barn protected

by an old Dutch design.
All around the outside skin
of our house, spiders pack up

their egg sacks, anchor
the next generation
snug in tight corners,

along ledges,
under flat stones.
In spring, the golden

FALL

spiderlings will hatch,
loose newly-spun lines,
catch the next silky gust out—

bet everything they have
that there's a place to anchor
on the other side.

SYLVIA ROSS

Guns and Roses

When I was a wee tiny girl, three or four times a year my mother would take my baby sister and me up into the Central Valley. We would drive a long way from our apartment building in Culver City. We would go into what I knew as "rich people's country," to my mother's aunt's home in Fresno.

Auntie had a house on East Thesta Street. It was the prettiest house I'd ever been in.

My great-aunt, the daughter of a cowboy and a Chukchansi Indian woman, had had the good fortune of being born beautiful. Although her early life was tragic, she survived. She married white and married well.

There was a rocking chair in Auntie's parlor just my size. She had big trees for the hot valley summer, and in the fall, fat black grapes hung down from the arbor in her backyard. Also in the fall, the roses along the front walk to her house recovered from the sere of summer. No longer heat-faded, they would show color as rich as they did in spring. Even in town, the world was quieter in 1941 than it is now. It was peaceful at my great-aunt's house, quiet enough that when leaves from the sycamore trees that lined her street fluttered down to the grass, they made a sound. The days were warm and dry. The light of fall afternoons was gentle. It was the best part of my day. My toddler sister would nap and I'd be free of the task of watching her. I could do nothing, and I liked doing nothing.

One afternoon I was sitting on the cool cement of the bungalow porch just watching the street. I could hear my mother and great-aunt's voices come through the window. I liked to listen to my mother when she was talking, especially if she didn't know I was there. I could hear Auntie trying to persuade my mother to come with her on a hunting trip over on the West Side. I knew that "West Side" meant the Coast Range hills. Though I couldn't yet read, my mother had taught me the landscape of her own years, and I knew the valley and the names of its surrounding mountains as well as I knew our apartment was on Venice Boulevard near Bagley Avenue.

"Only be gone four or five days. Charlie can't get off work," Auntie said, "but he can get the truck to drive us in an' come pick us up. Sister and I'll get two bucks sure. If you come, might be three or four."

Charlie was a relative Auntie talked about a lot. But I didn't know him. Charlie shot mountain lions for the government.

I wanted my mother to say "yes," but she didn't. I kept quiet as they talked about guns and shells and shot; I thought maybe I could go along if my mother would just say "yes." I wanted to meet an uncle who shot mountain lions, to climb on trails and shoot a deer. There was a picture of my great-aunt and her sister with their rifles in the room where my sister and I slept. In the picture my great-aunts wore shirts and pants like men, and high boots. At their feet were four antlered deer, heads lolling all twisted over, eyes wide open. The deer in the picture were my introduction to what it was to be dead. It seemed a very natural thing. My mother had taken me to a theater to see *Gone with the Wind*, but I hadn't seen *Bambi*. My consciousness had not been raised by cartoons.

My mother had told me that the women of my family—my mother, Auntie, and Auntie's sister—were Indians and we were not like other women. We did what we wanted. Observation had taught me that this was true. The women we knew had no children. In our building women played a card game called "bridge" on sleepy afternoons while their husbands were at work. Playing cards all afternoon seemed a very dull thing to do. I had gone with my mother when she was invited to a neighbor's apartment to play bridge. We didn't stay long. Mother found playing cards dull too. I liked that Auntie and her sister and my mother were women who took sage and salt, knives and rifles, and went hunting wearing shirts and pants like men.

Auntie continued to coax my mother to come along on the West Side. I heard my mother refer to little distractions like a pregnancy and a husband who wouldn't like her "going off." I wasn't sure what pregnancy meant, but thought "going off" was what we were doing when we came to visit Auntie. Hearing disappointment in my mother's voice, I knew she wouldn't agree to go. I wanted to show myself and argue in Auntie's favor. I wanted to wheedle and whine. With the knowledge of a child's power, I thought perhaps I could persuade Mother better than Auntie could. But I knew I couldn't say anything. I huddled down below the window even farther and kept very still. I'd learned long ago that discretion was the price I had to pay for eavesdropping.

We didn't get to go to the West Side and bring home venison with Auntie and her sister. I didn't learn to shoot at a living target, to gut and clean a carcass, or the joy of cooking fresh kill over an open fire.

The fogs of winter settled over the valley. Not until May did my mother drive us back up to the Central Valley. I had a baby brother by then. The roses in my great-aunt's yard were gorgeous and huge. In the dining room a big bouquet of the roses filled a big blue bowl in the center of the table. We followed Auntie through the house to the kitchen where she made a tea for my mother. Mother sat on a kitchen chair to nurse my brother. My little sister toddled to the pots and pan drawer and began to root around. Disassembled gun parts were scattered on the table and the room was filled with the good, strong smell of gun solvent. I asked my great-aunt if that was the gun she used when she went deer hunting. I know I used a sulky voice. I hadn't forgotten my disappointment in the fall and felt compelled to show it.

She laughed at me. "This is a sawed-off shotgun," she said. "This is for snakes, not for deer."

I was flippant and said, "There aren't any snakes here."

Auntie smiled at my mother and then turned to say to me, "Now, girl, a woman never knows when there might be a snake on the porch." This made my mother laugh, but I didn't think it very funny.

"Ellen," my great-aunt said to my mother as Baby Brother contentedly sucked away, "you have to teach this girl the difference between her guns. She's near five years old and she don't know a shotgun from a rifle." Auntie left the room for a minute and came back. She draped a big piece of buckskin across my shoulders from the tan bundle in her arms. It was heavy and soft. My little sister pulled out two pot lids and they clanged on the linoleum. Auntie showed my mother the other, larger skins she had tanned from the fall hunt. I put a finger through a hole where a bullet had gone through the skin. I asked what kind of a bullet it was and my mother said it was a .30-30, just like her rifle used.

    I knew Auntie would let me take the skin home to keep. The scent of the newly tanned skin was sharp compared to the fragrance of the roses drifting in from the dining room. I was yet to learn that while we were hunters we were not gatherers. It was Uncle Lou who tended the roses and brought them into the house. Deer skin and the roses' scent blended with odor of gun solvent in my great-aunt's kitchen. Auntie's kitchen was a good place to be. Baby Joe fell asleep but mother didn't put him down. My sister Nancy played with the pots and pans. My mother and Auntie talked in low voices all afternoon. Eventually Auntie finished cleaning the shotgun and put it away along with her cleaning kit. She washed up and began fixing supper. Her husband came home from work and I knew he was happy to see us.

    I was glad that the women of my family weren't like other women.

GEORGIANA VALOYCE-SANCHEZ

The Eye of the Flute

Enter the eye

From the north
a ribbon of geese drifts
high above the earth

   far below

beside a weathered wood shack
in the spring green foothills
of distant blue mountains
an old man sits
polishing stone

   Dogs bark in the distance

Down the hill a brown horse
black mane flying
runs along a reservation road
and three children and their mother
stand beside a fence
watching

   Beyond the fence a rusty tractor

sits fallow in the field
silent as the man beside it
watching the horse
run free

The eye watches   the eye
sees   the image
held
to still-point

Silence

Silence that holds all songs
that holds the breath
to play all songs
to life

hush

listen to the music

The horse is running still
hoofbeats on pavement a drum
black mane flying
free
within the still-point
of the song
   the locus of the poem
the eye of the flute

Three children and their mother
stand beside a fence
their father close by
all watching
the horse run free

Dogs back in the distance

The Eye of the Flute

An old man holds a polished stone
up to the sun
turning it
to catch the light

High above the earth
a ribbon of geese drifts south
the call of a long journey
echoing
across the endless sky

JANICE GOULD

U.C. Mascot, 1959

Now begins the festival and rivalry of late fall,
the weird debauch and daring debacle
of frat-boy parties as students parade cold streets in mock
funerals, bearing on shoulders scrawny effigies of dead
defeated Indians cut from trees, where,
in the twilight, they had earlier been hung.

"Just dummies," laughs Dad. "Indians hung
or burned—that's just for fun." Every Fall
brings the Big Game against Stanford, where
young scholars let off steam before the debacle
they may face of failed exams. "You're dead
wrong," he says to Mom. "They don't mock

real, live Indians." Around U.C. campus mock
lynchings go on. Beneath porches we see hung
the scarecrow Natives with fake long braids, dead
from merry-making. On Bancroft Way one has fallen
undecorously to a lawn, a symbol of the debacle
that happened a generation ago in California's hills, where

Indians were strung up. (White folks having fun? Where
did they go, those Indian ghosts?) "Their kids perform mock
war dances, whooping, re-enacting scenes of a debacle
they let loose," chides Mom. "Meanwhile we hang
portraits of Presidents on school walls and never let fall
the old red, white, and blue. My dear brother is dead

U.C. Mascot, 1959

because he fought in a White man's war. How many dead
Indians do they need to feel okay? This whole thing wears
on my soul." In the dark car we go silent, and the Fall
night gets chillier. In yards, blazing bonfires mock
the stars that glow palely above. A thin moon hangs
in the west. I've never hear the word "debacle"

before and wonder what it means. "What's a debacle,
Mom?" I ask. "Oh, honey, it's a terrible and deadly
collapse, complete ruin." I've noticed how the hung
Indians have their heads slumped forward. They wear
old clothes, headbands with feathers, face paint, moccasins instead of boots. Little do we know this Fall

living Indians at Feather Falls leave tobacco to mark
that, indeed, we're still here, lungs full of indigenous air.

SHAUNNA OTEKA MCCOVEY

Embolus

It waited that Fall
In the creek that fed the river,

On the road that led into the valley,
It waited,

And none of us saw it coming.
We had no idea we would hear

The heartbeat stammer, the breath leave.
The maple leaves knew before we did.

They caught the first light and
Hurled themselves early to the ground

In mourning, wishing it would hurry past,
Leave us before the first freeze,

But it didn't.
It waited until we least expected,

Until we took everything for granted.
It came,

And we prayed that our collective artery
Would survive the blow,

Embolus

That sunlight, songs and acorns
Would keep you strong,

Keep it at bay.
But it came,

And left us reeling and broken
With only the memory of your voice

To comfort us,
To guide us through.

DARRYL WILSON

Splashes of Red
Autumn 1867, Tawutlamit Wusci*

The odor of scorched gunpowder
   filled the air
In the morning
   it lay in soft, blue clouds
        over the earth of my people
   in the high desert
South of Modoc, west of Paiute

As the Great Powers of the seasons of our world
   move the goose and the salmon and the deer to migrate
So, too, that awesome power moves
   my people to gather
        for the last time before the long winter
To talk and to plan
   for the future and for the future of the children
·And so it is

They gathered there in fear
   knowing the soldiers were tracking them
Yet they obeyed that great power
   and gathered as is the custom of the ages

---

* The Infernal Caverns near Likely, California, where my grandmother was born. At Infernal Caverns, September 1867, the military of the United States fell upon a gathering of my people and committed unpardonable crimes.

Splashes of Red: Autumn 1867, Tuwutlamit Wusci

They smelled the sweat of the horses
   and of *weet-la* (devils)
They heard the distant report
   of the rifle

Yet they listened to the life of nature
   moving all around them
And they gathered
   at tuwutlamit wusci they gathered
   ...for the last time before the long winter

They did not think
   about the Paiute woman
   who slept with the soldiers
   ...And they told her of the gathering at tuwutlamit wusci

   ...And they came on sweating horses
   with their rifles in their hands

Frightened
   young mother ran towards the safety of tuwutlamit wusci
   ...Too late

When she looked back
   there was blood in her tracks
But she felt no pain
   for the pain was not hers to bear
Quickly she took the cradleboard from her back
   and lay it in the sunshine of autumn

The blood in her tracks—
   the infant on her back
   ...shot once through the neck
   once, but forever

With her hands she dug a little grave
   in a frightened crevasse of the trembling mountain
   ...And dried her tears with the dust of the sweet earth

FALL

She placed the eternal bundle
   In that shallow effort
   covered it with stones and a little flower growing nearby
   …And she, in fear—and a broken heart
   cried, among the splashes of red, Autumn, 1867…

STEPHEN MEADOWS

Alejandro

The close dark grain
of this antique chair
built about the time
you were born
holds for a moment
your face and its lines
entwining like the years
in my grandfather's field
a transient memory of you
your walk distinctive and measured
along the road
for wine and supper
looking as they all said
like a bear's in the shade
of live oak
you worked with the best of them
and had times been different
you might not have labored for wages
you might have had children
your signature practiced as a youth
more beautiful than my own
your untutored thoughts all those years
like the breathing of oaks
the plummet and rustle
of an acorn
in a green quiet place

CONTRIBUTOR BIOS

Roberta Reyes Cordero (Chumash) is active in revitalization of indigenous ocean traditions and is a co-founder of Chumash Maritime Association. One of several builders of the traditional Chumash canoe *'Elye'wun*, (Swordfish), she continues to participate in Chumash community canoe events. With degrees from the University of Washington School of Music and the University of Washington School of Law, she is now a professional mediator and peacemaker.

Janice Gould (Konkow) grew up in Berkeley and graduated from the University of California, Berkeley, where she received degrees in linguistics (B.A.) and English (M.A.). She earned her Ph.D. at the University of New Mexico with a dissertation on the poetry of Muscogee writer Joy Harjo. Her poems have been published in numerous journals and anthologies and collected in two books: *Beneath My Heart* (Firebrand Press, 1990) and *Earthquake Weather* (University of Arizona Press, 1996), as well as the artbook/chapbook *Alphabet*. Janice's scholarship on Native American poetry can be found in *Speak to Me Words: Essays on American Indian Poetry*, which she co-edited with Dean Rader (University of Arizona Press, 2003). She works at the University of Arizona's Center for Creative Photography while studying for a master's degree in Library Science.

Ishi (Yahi) was born in the foothills near Mount Lassen. By the 1870s, most of the Yahi had been wiped out in a succession of massacres, and in 1911, a lone, starving Indian arrived in Oroville. City officials took the man to the U.C. anthropology museum in San Francisco. In accordance with Yahi custom, he never revealed his given name to anyone; anthropologist Alfred Kroeber called him "Ishi," the Yahi word for man. Ishi lived at the museum, assisting linguists and anthropologists and giving public demonstrations as well as working as a janitor. Ishi died in 1916 from tuberculosis.

Frank LaPena (Nomtipom Wintu), an internationally exhibited painter and published poet, was born in 1937 in San Francisco, California. As a young man he became interested in the song, dance, and ceremonial traditions of his tribe and other neighboring tribes. With the Maidu Dancers and Traditionalists, he has been active in the revival and preservation of these native arts. He is professor emeritus of art and ethnic studies at California State University, Sacramento. He is the author of *Dream Songs and Ceremony* (Heyday Books, 2004).

Contributor Bios

Shaunna Oteka McCovey (Yurok/Karuk) wrote her first poem at the age of six while growing up on the Yurok Indian reservation in northern California. She holds master's degrees in social work and environmental law and a juris doctorate from Vermont Law School. Her poems and essays have appeared in *News from Native California, Through the Eye of the Deer, The Dirt Is Red Here,* and *Eating Fire, Tasting Blood: Breaking the Great Silence of the American Indian Holocaust.* She is the author of a chapbook of poetry entitled *Swim You Every River,* and her first full-length book of poetry is called *The Smokehouse Boys.*

Stephen Meadows (Ohlone) holds degrees from the University of California, Santa Cruz, and San Francisco State University. He has been influenced and inspired by the poems of ancient China as well as the works of Dylan Thomas, Kenneth Rexroth, Gary Snyder, and the Beat poets of San Francisco. One of his poems is on a bronze plaque in San Francisco. He is a 20-year veteran of Public Radio and has interviewed scores of musicians and newsmakers from the British Isles to Canada to the United States. He resides in the foothills of the Sierra Nevada.

Deborah Miranda (Esselen) received her Ph.D. from the University of Washington and is currently associate professor of English at Washington and Lee University in Lexington, VA. For the academic year 2007-2008, Deborah will be IAC Visiting Scholar in Ethnic Studies with the American Indian Studies Center at UCLA. Her book *Indian Cartography* (Greenfield Review Press, 1999) won the North American Native Authors First Book Award, and she is also the author of *The Zen of La Llorona* (Salt Publishing, 2005).

Francisco Patencio (Kauisik Cahuilla) was born in Chino Canyon in 1856. He became a noted elder, religious leader, and storyteller, and his *Stories and Legends of the Palm Springs Indians* was published in 1943. He died in 1947.

Dorothy Ramon (Serrano) was born in 1909 on the Morongo Indian Reservation in Riverside County and is the author, with Eric Elliott, of *Wayta' Yawa': "Always Believe."* She passed away in 2002.

Sylvia Ross (Chukchansi) was born and raised in Los Angeles, apart from her family's culture. After graduating from high school, she worked as a painter for Walt Disney Productions. Once married, she returned to school and earned a B.A. from Fresno State University. For many years she taught at a school attended by the children of the Tule River Indian Reservation. Now retired, she and her husband live in the Sierra foothills near the town of Exeter. She is the author of *Lion Singer* (Heyday Books, 2005) and is a frequent contributor to *News from Native California*.

Contributor Bios

Greg Sarris (Coast Miwok) was born in 1952 and grew up in Santa Rosa. He is the author of several books, including *Grand Avenue* (Hyperion, 1994) and *Watermelon Nights* (Hyperion, 1998). He received his Ph.D. in Modern Thought and Literature from Stanford University. He holds the Federated Indians of Graton Rancheria Endowed Chair in Native American Studies in the School of Humanities at Sonoma State University, and is serving his seventh consecutive term as chairman of the Federated Indians of Graton Rancheria.

Lucy Thompson (Yurok) was born in Klamath River country, in the Yurok village of Pecwan. Although she lived much of her adult life in white society, she had an abiding passion for her own culture and was annoyed and appalled by the inaccuracies in accounts of Yurok ways written by white anthropologists. In 1916 she decided to set the record straight and published *To the American Indian*, a detailed and unique survey of Yurok culture.

Georgiana Valoyce-Sanchez (Chumash/Tohono-O'odham) is a storyteller, educator, and nationally published writer. She has taught for the American Indian Studies Program at California State University, Long Beach since 1987. She is also a board member of the California Indian Storytellers Association and is active in the preservation of California Indian languages, sacred sites, and traditional arts.

Darryl Wilson, or Sul'ma'ejote, was born in 1939 at the confluence of Sul'ma'ejote (Fall River) and It ajuma (Pit River), in northeastern California. He graduated from Enterprise High School, Redding, California. He holds a B.A. from UC Davis and an M.A. and Ph.D. from the University of Arizona, Tucson. He has seven sons and has been a grampa four times. Currently he lives in Carson City, Nevada.

PERMISSIONS

"Frost" by Greg Sarris; "New Year's Day" by Janice Gould; "Petroglyph" by Deborah Miranda; "Winter Beginning, Winter Ending" by Sylvia Ross; "Winter Comes" by Georgiana Valoyce-Sanchez from *News from Native California* Vol. 20, No. 2 (Winter 2006/2007). All selections © 2006 by their author and used with permission.

"Conception" by Shaunna Oteka McCovey; "In the Mountains" by Stephen Meadows; "Iris" by Greg Sarris; "*Kweme Psukitok*, Spring Maiden; *Amal*, Flower Maiden" by Darryl Wilson; "Water" by Deborah Miranda from *News from Native California* Vol. 20, No. 3 (Spring 2007). All selections © 2007 by their author and used with permission.

"Fireworks" by Shaunna Oteka McCovey; "In Praise of August" by Deborah Miranda; "In the Water over Stones" by Stephen Meadows; "Osprey" by Greg Sarris; "Summer 1945" by Georgiana Valoyce-Sanchez (reprinted from *The Sound of Rattles and Clappers: A Collection of New California Indian Writing*, ed. Greg Sarris, University of Arizona Press, 1994) from *News from Native California* Vol. 20, No. 4 (Summer 2007). All selections © 2007 by their author and used with permission.

"Embolus" by Shaunna Oteka McCovey; "Faith" by Deborah Miranda; "Guns and Roses" by Sylvia Ross; "U.C. Mascot, 1959" by Janice Gould from *News from Native California* Vol. 21, No. 1 (Fall 2007). All selections © 2007 by their author and used with permission.

"The Acorn" is from *To the American Indian: Reminiscences of a Yurok Woman* by Lucy Thompson, originally published 1916, reprinted by Heyday Books (1991).

"*Ascui* (When the Ice Cracks)" by Darryl Wilson © 2007 by Darryl Wilson. Reprinted by permission of the author.

"The Burial" and "Alejandro" by Stephen Meadows © 2007 by Stephen Meadows. Reprinted by permission of the author.

Permissions

Excerpt from "Coyote and His Sister" from *Yana Texts* by Edward Sapir, University of California Publications in American Archaeology and Ethnology Vol. 9, No. 1 (February 19, 1910).

Excerpt from "Coyote's Journey" by William Bright, *American Indian Cultural and Research Journal* 4 (1980).

"Coyote and the Moons" from "Shasta Myths" by Roland B. Dixon, *Journal of American Folk-Lore* Vol. 23, No. 87 (January–March, 1910).

"Coyote and Raccoon" from *The North American Indian* Vol. 13 by Edward S. Curtis, The University Press, 1924.

"The Eye of the Flute" by Georgiana Valoyce-Sanchez (reprinted from *The Sound of Rattles and Clappers: New California Indian Writing* ed. Greg Sarris, University of Arizona Press) © 1994 by Georgiana Valoyce-Sanchez. Reprinted by permission of the author.

Excerpt from "Four Dream Cult Songs" from *The 1870 Ghost Dance* by Cora Du Bois, *Anthropological Records* Vol. 3, No. 1 (1939).

"Haiku/Senryu" by Roberta Cordero © 2007 by Roberta Cordero. Reprinted by permission of the author.

Excerpt from *Ishi's Tale of Lizard* trans. Leanne Hinton (Farrar, Straus & Giroux, 1992) © 1992 by Leanne Hinton. Reprinted by permission of the translator.

"The Lazy Man and the Tamciye" from *Atsugewi Ethnography* by Thomas R. Garth, *Anthropological Records* Vol. 14, No. 2 (Feb. 27, 1953).

"Picking Yucca Flowers" and "Why People Did Not Kill Tarantulas" from *Wayta' Yawa': "Always Believe"* by Dorothy Ramon and Eric Elliott (Malki Museum Press, 2000) © 2000 by Malki Museum Press. Reprinted by permission of Malki Museum Press.

"Puimeminmak, the Deer Maker" from *Wintu Myths* by Cora Du Bois and Dorothy Demetracopoulou, University of California Publications in American Archaeology and Ethnology Vol. 28, No. 5 (Dec. 29, 1931).

"The Quail Legend" from *Desert Hours with Chief Patencio* by Francisco Patencio (Palm Springs Desert Museum, 1971).

Permissions

"Rattlesnake Ceremony Song" from *Handbook of the Indians of California* by Alfred Kroeber, U.S. Government Printing Office, 1925.

"Splashes of Red" by Darryl Wilson (reprinted from *The Sound of Rattles and Clappers: New California Indian Writing* ed. Greg Sarris, University of Arizona Press) © 1994 by Darryl Wilson. Reprinted by permission of the author.

"Spring Dance" from *Nomlaki Ethnography* by Walter Goldschmidt, University of California Publications in American Archaeology and Ethnology Vol. 42, No. 4 (May 22, 1951).

"The Wolf Makes the Snow Cold" from *Maidu Myths* by Roland B. Dixon, *Bulletin of the American Museum of Natural History* Vol. 17, Part 2 (June 30, 1902).